T0173357

# Primary **Maths** for **Scotland**

# 2nd Level Maths
## Textbook 2C

**Series Editor:** Craig Lowther

**Authors:** Antoinette Irwin, Carol Lyon,
Kirsten Mackay, Felicity Martin, Scott Morrow

# Contents

# 1 Estimation and rounding

## 1.1 Rounding whole numbers

We are learning to round whole numbers to the nearest 1000, 10000 or 100000.

**Before we start**

What is the biggest number you can make using the digits 2, 5, 6, 7, 8 and 9? Explain what each digit represents. What is the smallest number that can be made with these digits? Again, explain what value each digit represents in your new number.

We can use an empty number line to help us decide whether to round up or round down.

**Let's learn**

To round 523871 to the nearest 1000 we look at multiples of 1000 on either side. 523871 is between 523000 and 524000, but is closer to the number 524000 so we round up to 524000.

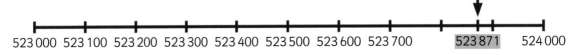

523000 523100 523200 523300 523400 523500 523600 523700 **523871** 524000

To round 523871 to the nearest 10000 we look at multiples of 10000 on either side. 523871 is between 520000 and 530000, but is closer to the number 520000 so we round down to 520000.

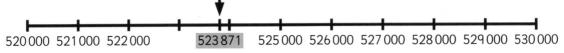

520000 521000 522000 **523871** 525000 526000 527000 528000 529000 530000

To round 523871 to the nearest 100000 we look at multiples of 100000 on either side. 523871 is between 500000 and 600000, but is closer to the number 500000 so we round down to 500000.

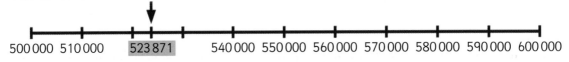

500000 510000 **523871** 540000 550000 560000 570000 580000 590000 600000

**Let's practise**

1) Draw an empty number line and write on the two multiples of 10 000 that are either side of each number. Then use this to estimate where the actual number lies and round to the nearest 10 000.

   a)  478 939 lies between [ ] and [ ] rounded [ ]

   b)  847 235 lies between [ ] and [ ] rounded [ ]

   c)  765 812 lies between [ ] and [ ] rounded [ ]

   d)  38 036 lies between [ ] and [ ] rounded [ ]

2) Round these numbers to the nearest 100 000:

   a)  598 388        b)  780 623        c)  384 003
   d)  180 763        e)  839 062        f)  238 929
   Draw an empty number line to help if required.

3) Nuria has been working on rounding numbers to the nearest 10 000 and 100 000. Is she correct? If not, what should her answer be?

   a)  385 835 rounded to the nearest 10 000 is 380 000.
   b)  579 889 rounded to the nearest 100 000 is 600 000.
   c)  84 890 rounded to the nearest 10 000 is 100 000.
   d)  449 837 rounded to the nearest 100 000 is 500 000.
   e)  113 987 rounded to the nearest 10 000 is 110 000.
   f)  473 950 rounded to the nearest 10 000 is 460 000.
   g)  807 999 rounded to the nearest 100 000 is 800 000.
   h)  709 014 rounded to the nearest 10 000 is 70 000.

**CHALLENGE!**

Work with a partner. You will need a set of numeral cards from 0 to 9. Shuffle the cards and lay them out face down. Take it in turns to pick six numeral cards and make a six-digit number of your choice. Ask your partner to round the number to nearest 1000, 10 000 and 100 000 for the number you chose. Check their answers are correct! Then swap roles.

# 1 Estimation and rounding

## 1.2 Rounding decimal fractions

> We are learning to round decimal fractions to the nearest $\frac{1}{100}$.

**Before we start**

What is this number? **7.41**

What would this number be rounded to the nearest $\frac{1}{10}$?

What about rounded to the nearest whole number?

> We can round decimal fractions up or down to the nearest $\frac{1}{100}$.

**Let's learn**

When we round decimal fractions to the nearest $\frac{1}{100}$ we look at the third digit $\left(\frac{1}{1000}\right)$ after the decimal point.

If this digit is 5 or more we usually round up to the nearest $\frac{1}{100}$.

If this digit is less than 5 we usually round down to the nearest $\frac{1}{100}$.

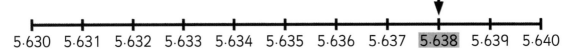

5.630  5.631  5.632  5.633  5.634  5.635  5.636  5.637  5.638  5.639  5.640

5.638 is between 5.63 and 5.64, but is closer to 5.64 so we round up to 5.64.

**Let's practise**

1) Round each of these decimal fractions to the nearest hundredth:

    a) 5·872        b) 18·467       c) 7·809

    d) 0·012        e) 4·998        f) 0·939

2) Are these number statements true or false? Circle true or false.

    a) 9·312 rounded to the nearest hundredth is 9·32    **true/false**

    b) 5·955 rounded to the nearest hundredth is 5·95    **true/false**

    c) 0·048 rounded to the nearest hundredth is 0·05    **true/false**

    d) 14·715 rounded to the nearest hundredth is 14·7    **true/false**

    e) 9·028 rounded to the nearest hundredth is 9·08    **true/false**

3) Can you round these measurements to two decimal places?

    a) 6·982 kg     b) 793·872 km    c) 0·019 litres    d) 8·008 kg

**CHALLENGE!**

Amman walks 2·346 km on Saturday, 1·784 km on Sunday and 0·928 km on Monday. Round each distance to the nearest hundredth to estimate the total distance he has walked over the three days.

If you added the exact amounts and then rounded the answer to the nearest hundredth, would you get a different answer?

What does this tell you about how accurate your estimation was?

## 1.3 Using rounding to estimate the answer

We are learning to apply our knowledge of rounding to estimate the answer to problems.

**Before we start**

Isla says that the number 387 097 is 387 000 rounded to the nearest 10 000. Is she correct? Explain your answer.

When you solve a problem, it is always a good idea to check your answer. We can estimate the answer to a problem using rounding to check if the answer given is reasonable.

**Let's learn**

Finlay is working out 8294 − 4372 and gets the answer 7822. Amman is checking the answers. He thinks this answer is wrong.

He quickly estimates the answer using rounding:

8294 rounded to the nearest 1000 is 8000 and 4372 rounded to the nearest 1000 is 4000. He can see straight away that the answer should be approximately 8000 − 4000, which is 4000.

7822 is not a reasonable answer, so he asks Finlay to try again.

We can use rounding to estimate the answer to multiplication, too.

Isla and Nuria are working out 62 × 8.

Isla's answer is 208. Nuria's answer is 496.

We can estimate the answer using rounding to check who is correct.

If we round 62 to 60, we can estimate the answer should be around 480.

6 × 8 = 48, so 60 × 8 = 480.

This shows us that Nuria's answer is very close to the estimate, so her answer is more reasonable.

**Let's practise**

1) Say whether each answer is reasonable or not. Explain your answer.

| | Reasonable? | |
|---|---|---|
| | Yes | No |
| Dana multiplied 72 by 9 and got 648 | | |
| Hamad multiplied 46 by 6 and got 276 | | |
| Jameela multiplied 57 by 7 and got 599 | | |

2) Isla has worked out these problems as below. Check her answers by rounding and estimating and decide if her answers are reasonable. Explain your thinking for each one.

a) $14\,987 + 44\,537 = 19\,524$

b) $89\,765 - 5273 = 84\,492$

c) $45 \times 19 = 405$

d) $4872 - 3928 = 8800$

e) $35 \times 21 = 735$

f) $387 + 268 + 1938 = 2053$

g) $379\,928 - 217\,304 = 108\,624$

h) $81 \times 79 = 6399$

3) The school fair for charity raises a total of £2413. Estimate how much each charity would get if this total was split between:

a) three charities    b) four charities        c) ten charities

Explain your answers.

**CHALLENGE!**

Play a game of Range Finder with a partner. Place a counter each on the START square.

Take turns to spin a spinner (1–6) and move that number of spaces, horizontally or vertically. If you land on a number you must estimate, by rounding, the product of that number multiplied by the spinner number. Place a cube on the table below, in the range of your estimate. Now, calculate the answer. If your estimate was correct, move your cube to cover the number.

| 403 | | 888 | | 197 | |
|---|---|---|---|---|---|
| | 649 | | 707 | | 937 |
| 621 | | 817 | | 588 | |
| | 233 | | 362 | | 516 |
| 741 | | 496 | | 924 | |
| START | 277 | | 536 | | 356 |

The player who gets three cubes in a row, horizontally, vertically or diagonally, is the winner.

| Below 500 | 500 to 1000 | 1000 to 1500 | 1500 to 2000 | 2000 to 2500 | 2500 to 3000 | 3000 to 3500 | Over 3500 |
|---|---|---|---|---|---|---|---|

## 2.1 Reading and writing whole numbers

> We are learning to read and write whole numbers with six or more digits.

### Before we start

Using the numerals below only once each time, **make three different five-digit numbers** with:

a) 0 in the hundreds place and 6 in the tens place

b) 2 in the tens of thousands place and 8 in the ones place

Write each number in both numerals and words.

**6   3   2   8   0**

> The hundreds, tens and ones pattern can help us read and write big numbers.

### Let's learn

Isla uses the numerals 2, 6, 5, 7, 0 and 4 to make a six-digit number.

She writes her number in words and in numerals.

Isla puts a comma after the word 'thousand' and writes 'and' before the tens and ones words. Her number says **two hundred and sixty-five thousand, seven hundred and four**.

| Thousands | | | Ones | | |
|---|---|---|---|---|---|
| H | T | O | H | T | O |
| 2 | 6 | 5 | 7 | 0 | 4 |

When writing her number in numerals, she leaves a small space between the thousands digit and the hundreds digit. For example, **265 704**.

**Let's practise**

1) Write these six-digit numbers in words.

    a) 122 913        b) 808 708        c) 400 026

    d) 167 090        e) 300 001        f) 999 999

2) Write these six-digit numbers in numerals.

    a) four hundred and three thousand, five hundred and sixty-one

    b) three hundred and ninety-nine thousand, two hundred and forty

    c) two hundred thousand and six

    d) seven hundred and twenty thousand and eighteen

    e) nine hundred and seventy-two thousand, nine hundred and eleven

    f) five hundred thousand, two hundred

3) Each of these coloured strips represents a six-digit number. Here is the key for each digit:

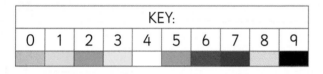

| KEY: | | | | | | | | | |
|---|---|---|---|---|---|---|---|---|---|
| 0 | 1 | 2 | 3 | 4 | 5 | 6 | 7 | 8 | 9 |

For each coloured strip below, write the number it represents in numerals and then in words.

a)       b)       c)

---

**CHALLENGE!**

These place value houses show a **seven-digit number**.

In numerals: **1 116 510**

In words: **one million, one hundred and sixteen thousand, five hundred and ten**.

| Millions | | | Thousands | | | Ones | | |
|---|---|---|---|---|---|---|---|---|
| H | T | O | H | T | O | H | T | O |
| | | 1 | 1 | 1 | 6 | 5 | 1 | 0 |

Can you write the number shown by these place value houses in both words and numerals?

| Millions | | | Thousands | | | Ones | | |
|---|---|---|---|---|---|---|---|---|
| H | T | O | H | T | O | H | T | O |
| | | 8 | 2 | 3 | 6 | 4 | 7 | 7 |

# 2 Number – order and place value

## 2.2 Representing and describing whole numbers

We are learning to build and describe six-digit whole numbers.

**Before we start**

Using the digits on these cards, write a number to match each clue.

a) The largest five-digit number where the value of the 3 is 30 000.

b) The smallest five-digit number with zero in the hundreds place.

c) An odd number with ninety-seven thousands.

The position of a digit tells us its value. Each place to the left is worth 10 times more.

**Let's learn**

To find the value of a digit, we multiply it by its place. For example, 167 935:

The 1 is worth 1 × 100 000 = 100 000

The 6 is worth 6 × 10 000 = 60 000

The 7 is worth 7 × 1000 = 7000

The 9 is worth 9 × 100 = 900

The 3 is worth 3 × 10 = 30

The 5 is worth 5 × 1 = 5

| Thousands | | | Ones | | |
|---|---|---|---|---|---|
| H | T | O | H | T | O |
| 1 | 6 | 7 | 9 | 3 | 5 |

**167 935 = 100 000 + 60 000 + 7000 + 900 + 30 + 5**

## Let's practise

1) Write the value of the blue digit in both words and numerals. For example, **7**11 238 – seven hundred thousand and 700 000.

   a) 2**4**3 309     b) 5**2**4 700     c) **6**78 026     d) 900 40**1**
   e) **3**33 333     f) 814 814     g) 457 0**3**2     h) 1**8**0 217

2) Nuria made the number **411 236** with place value arrow cards.

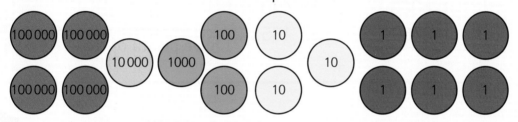

Isla made the same number with place value counters.

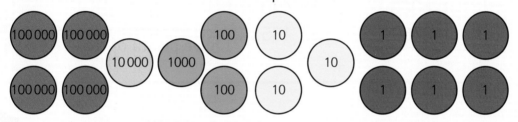

Represent the numbers below using place value arrow cards or place value counters. Write each number in numerals and draw the cards or counters you used each time.

   a) five hundred thousand, two hundred and four
   b) one hundred thousand and nine
   c) six hundred and twenty thousand, one hundred
   d) two hundred and seventeen thousand and forty

3) Use place value arrow cards or place value counters to represent a six-digit number of your choice. Ask a partner to write your number in numerals and in words. Are they correct?

☆ **CHALLENGE!**

Nuria is thinking of a six-digit number. Three of the digits are 2s. There is an odd number in the tens place and a zero in the ones place. The sum of the digits is 8. What could Nuria's number be?

## 2.3 Place value partitioning of whole numbers

We are learning to partition six-digit numbers in different ways.

**Before we start**

Amman wants to make four-digit numbers using base 10 blocks, but the thousand blocks have gone missing! How could he make the following numbers using only hundreds, tens and ones blocks?

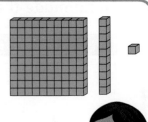

a) 6000          b) 4903

c) 2760          d) 1358

I have plenty of these.

Being able to partition numbers in different ways helps us to calculate efficiently.

**Let's learn**

Let's partition the number 435 218.

There are 435 thousands. Their value is 435 000.

| Thousands | | | Ones | | |
|---|---|---|---|---|---|
| H | T | O | H | T | O |
| 4 | 3 | 5 | 2 | 1 | 8 |

×10  ×10  ×10  ×10

There are two hundreds. Their value is 200.

There is one ten. Its value is 10.

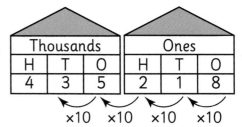

Each place to the left is worth 10 times more.

There are eight ones. Their value is 8.

435 thousands + 2 hundreds + 1 ten + 8 ones = 435 218

435 000    +    200    + 10 +  8    = 435 218

## Let's practise

1) Write the number represented by these place value houses in four different ways. One has been done for you.

| Thousands | | | Ones | | |
|---|---|---|---|---|---|
| H | T | O | H | T | O |
| 3 | 1 | 7 | 4 | 2 | 4 |

317 thousands, 4 hundreds, 2 tens and 4 ones

317 thousands, 4 hundreds, 24 ones

317 thousands, 42 tens and 4 ones

317 thousands and 424 ones

a)

| Thousands | | | Ones | | |
|---|---|---|---|---|---|
| H | T | O | H | T | O |
| 7 | 9 | 8 | 8 | 5 | 1 |

b)

| Thousands | | | Ones | | |
|---|---|---|---|---|---|
| H | T | O | H | T | O |
| 4 | 0 | 4 | 9 | 6 | 9 |

c)

| Thousands | | | Ones | | |
|---|---|---|---|---|---|
| H | T | O | H | T | O |
| 2 | 6 | 0 | 5 | 8 | 6 |

d)

| Thousands | | | Ones | | |
|---|---|---|---|---|---|
| H | T | O | H | T | O |
| 4 | 3 | 5 | 2 | 1 | 8 |

2) The number 456 219 has been partitioned in six different ways (and more partitions are possible!).

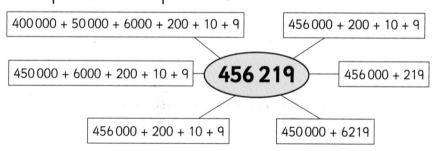

400 000 + 50 000 + 6000 + 200 + 10 + 9

456 000 + 200 + 10 + 9

450 000 + 6000 + 200 + 10 + 9

**456 219**

456 000 + 219

456 000 + 200 + 10 + 9

450 000 + 6219

Work with a partner to find six different ways to partition each of these numbers.

a) 210 348

b) 763 234

c) 653 270

d) 342 091

e) 515 200

f) 109 867

## CHALLENGE!

Amman writes 92 000 + 1000 + 700 + 10 + 3. Explain why Amman is incorrect. Partition the number 921 713 correctly. Can you find more than one solution?

Partition the number 921 713.

## 2.4 Number sequences

We are learning to sequence numbers using the patterns in our place value system.

**Before we start**

Finlay's pen has leaked all over his work! Can you identify the missing numbers?

| 14 013 | | | 14 313 | 14 413 | | 14 613 | 14 713 | | 14 913 |
|---|---|---|---|---|---|---|---|---|---|
| | 15 113 | 15 213 | 15 313 | | 15 513 | | 15 713 | 15 813 | 15 913 |
| | | | | 16 413 | 16 513 | | | 16 813 | |

The repeating pattern in our place value system can help us to sequence numbers.

**Let's learn**

Grouping the digits into ones and thousands can help us to work out the number that comes after 439 999. Think about the 'ones house' first. 999 + 1 = 1000. Next think about the 'thousands house'.
439 000 + 1000 = 440 000.
So the number after 439 999 is 440 000.

| Thousands | | | Ones | | |
|---|---|---|---|---|---|
| H | T | O | H | T | O |
| 4 | 3 | 9 | 9 | 9 | 9 |

**Let's practise**

1) Write the next five numbers in each sequence.

   a) 218 376, 218 377, 218 378 …   b) 779 996, 779 997, 779 998 …
   c) 800 007, 800 008, 800 009 …   d) 611 595, 611 596, 611 597…
   e) 304 997, 304 998, 304 999 …   f) 199 996, 199 997, 199 998 …

2) Now write the next five numbers in each of these sequences.

   a) 321 504, 321 503, 321 502 …   b) 578 236, 578 235, 578 234 …
   c) 999 999, 999 998, 999 997 …   d) 610 103, 610 102, 610 101 …
   e) 128 004, 128 003, 128 002 …   f) 400 006, 400 005, 400 004 …

3) Look at these number sequences. Are the numbers increasing or decreasing? How much bigger or smaller are they getting each time?

   a) 420 003, 520 003, 620 003, 720 003
   b) 789 619, 789 519, 789 419, 789 319
   c) 190 145, 200 145, 210 145, 220 145
   d) 800 019, 800 009, 799 999, 799 989

Increasing means getting bigger. Decreasing means getting smaller.

**CHALLENGE!**

We know that:

10 × 1 = 10
10 × 10 = 100
10 × 100 = 1000
10 × 1000 = 10 000
10 × 10 000 = 100 000
10 × 100 000 = one million
10 × 100 000 = 1 000 000

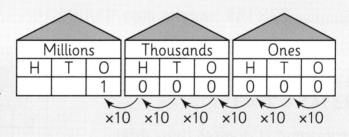

1 million is a seven-digit number.

Read these numbers with a partner and work together to write the next five numbers in each sequence.

   a) 1 000 001, 1 000 002, 1 000 003 …
   b) 5 000 000, 6 000 000, 7 000 000 …
   c) 2 000 400, 2 000 500, 2 000 600 …

## 2.5 Comparing and ordering whole numbers

We are learning to compare and order whole numbers with up to six digits.

**Before we start**

a) Use the numbers inside the ovals. How many different ways can you find to make each statement true?

$\boxed{\phantom{00}} < \boxed{\phantom{00}}$

$\boxed{\phantom{00}} > \boxed{\phantom{00}}$

b) Now write the numbers in order from smallest to largest.

$(17\,009)$ $(19\,700)$ $(17\,900)$ $(19\,070)$

To compare and order numbers, we need to think carefully about the values of the digits.

**Let's learn**

### Comparing numbers

Let's **compare** the numbers **273 141** and **274 311**.

Both numbers have the digit 2 in the hundreds of thousands place and the digit 7 in the tens of thousands place.

The number 273 141 has the digit 3 in the thousands place.

The number 274 311 has the digit 4 in the thousands place.

Three thousand is less than four thousand, so **273 141 < 274 311**.

Four thousand is greater than three thousand, so **274 311 > 273 141**.

| Thousands | | | Ones | | |
|---|---|---|---|---|---|
| H | T | O | H | T | O |
| 2 | 7 | 3 | 1 | 4 | 1 |
| 2 | 7 | 4 | 3 | 1 | 1 |

**Let's practise**

1) Copy and complete each statement. Write < or > for each box.
   Read your answers aloud to a partner.
   a) 230 290 ☐ 290 230
   b) 100 389 ☐ 10 389
   c) 347 133 ☐ 347 313
   d) 900 099 ☐ 900 909
   e) 789 572 ☐ 879 572
   f) 400 005 ☐ 40 005
   g) 665 504 ☐ 656 540
   h) 575 541 ☐ 575 543

2) Write true or false for each statement. Change some words in the
   false statements to make them true.
   a) 804 600 = eighty-four thousand, six hundred
   b) 500 090 < fifty thousand and ninety
   c) 113 311 > one hundred and thirty-one thousand, one hundred
      and thirteen
   d) 487 650 > four hundred and eighty-seven thousand, five hundred
      and sixty
   e) 355 007 < three hundred and seventy-seven thousand and five

3) a) Nuria rolls a dice six times to make the number five hundred and
      forty-two thousand, one hundred and thirty-five.

         **542 135**

   She challenges Finlay to write **two multiples of 10**, one that is
   smaller than 542 135 and one that is larger than 542 135. What
   could Finlay's numbers be? Write three pairs of possible answers.

   b) Work with a partner. Take turns to roll a dice six times to make
      a six-digit number. Challenge your partner to write two six-digit
      numbers, one that is smaller than your number and one that is
      larger. Increase the challenge by creating a rule, for example,
      'Both numbers must have zero in the thousands place'.

## Let's learn

### Ordering numbers

Look at how many digits each number has.

12 380      120 380      120 830      130 208      103 008

The number 12 380 has five digits. All of the other numbers have six digits, so **12 380** is the smallest number.

Next, compare all of the six-digit numbers from left to right.

All have 1 in the hundred thousands place so we need to look at the ten thousands place.

**103 008** has the digit 0 in the ten thousands place so it is the second smallest number.

Two numbers have the digit 2 in the ten thousands place. **120 380** is smaller than **120 830**.

The number **130 208** has the digit 3 in the tens of thousands place. It is the largest number.

From smallest to largest the correct order is:

12 380      103 008      120 380      120 830      130 208

## Let's practise

4) Write each set of numbers in ascending order (smallest to largest).
   a) 233 421      233 419      233 424      233 428      233 417
   b) 489 667      489 761      489 656      489 682      489 759
   c) 944 009      949 016      941 010      944 023      944 020

5) This table shows the diameter of the planets in our Solar System. Write the names of the planets in ascending size (smallest to largest).

| Planet | Diameter in km |
|--------|----------------|
| Mercury | 4876 |
| Venus | 12107 |
| Earth | 12755 |
| Mars | 6794 |
| Jupiter | 142983 |
| Saturn | 120536 |
| Uranus | 51117 |
| Neptune | 49527 |
| Pluto | 2390 |

6) a) Use the numerals on the cards to make different six-digit numbers that fit the criteria.

2   1
8   0
5   9

- an even number
- a multiple of 5
- an odd number
- a multiple of 10
- the largest possible number
- a multiple of 25

b) Now put your numbers in descending order (largest to smallest).

**CHALLENGE!**

a) Use the cards below to make nine different six-digit numbers.

b) Write six different statements about your numbers using the symbols < or >.

c) Write all nine numbers in ascending or descending order. Tell your teacher which order you have chosen.

| Two hundred and twenty-one thousand... | ...six hundred and forty-five |
| One hundred and twelve thousand... | ...six hundred and fifty-four |
| Two hundred and twelve thousand... | ...five hundred and fifty-six |

## 2.6 Negative numbers

> We are learning to solve problems involving negative numbers.

**Before we start**

Write the temperature shown by these number lines in both words and numerals.

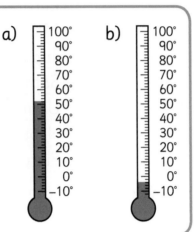

a)    b)

> A number line can help us solve problems involving positive and negative numbers.

**Let's learn**

Number lines can be horizontal, vertical or circular.

The circular number line shows a temperature of **23°C**.

The average daily temperature during the month of February was negative three degrees Celsius (–3°C).

Amman uses a positive and negative number line to work out the difference between the two temperatures.

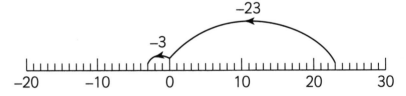

> A big jump of 23 will get me to zero.
> I need to jump back three more spaces to land on negative 3.
> 23 add 3 makes 26 so the difference is 26°C.

**Let's practise**

These tables show the average December temperature in places around the world.

Use a number line to help you solve these problems.

| Amsterdam | 4°C |
|---|---|
| Athens | 12°C |
| Berlin | 2°C |
| Bucharest | 0°C |

| Cairo | 24°C |
|---|---|
| Glasgow | 5°C |
| Helsinki | –4°C |
| Moscow | –9°C |

| North Pole | –13°C |
|---|---|
| Orlando | 22°C |
| Reykjavik | –1°C |
| Sydney | 26°C |

1) In which places is the average December temperature below freezing?

2) How many degrees warmer is the average December temperature in Orlando than in Bucharest?

3) How many degrees colder is it in Helsinki than in Glasgow?

4) In which place is the average December temperature 11°C colder than in Berlin?

5) In which place is the average December temperature 27°C warmer than in Reykyavik?

6) What is the average difference in temperature between Athens and Helsinki?

7) What is the average difference in temperature between the coldest and hottest places?

**CHALLENGE!**

Arrange these dive depths in ascending order.

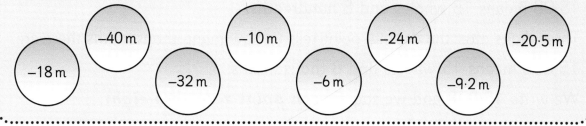

–18 m    –40 m    –32 m    –10 m    –6 m    –24 m    –9·2 m    –20·5 m

# 2 Number – order and place value

## 2.7 Reading and writing decimal fractions

We are learning to read and write decimal fractions with thousandths.

### Before we start

The children agree that this diagram shows *one and six hundredths* but disagree about how to read and write this as a decimal fraction. Who is correct? Explain.

We say one point zero six.

No! It's one point six. We can forget about the zero.

Write one, then a decimal point, then zero, then six.

Just write 6 after the decimal point.

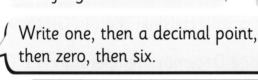
The three digits after the decimal point tell us how many thousandths there are.

### Let's learn

The digits in a decimal fraction have place values, just like those in whole numbers.

The digit to the right of the decimal point tells us how many tenths there are. 15·8 means 15 wholes and 8 tenths.

Two digits after the decimal point tells us how many hundredths there are.

15·08 means 15 wholes and 8 hundredths.

Three digits after the decimal point tells us how many thousandths there are.

15·008 means 15 wholes and 8 thousandths.

We write **15·008** and we say *fifteen point zero zero eight.*

**Let's practise**

1) How would you say these decimal fractions? One has been done for you.

   a) **4·736 → four point seven three six**
   b) 18·421      c) 37·034      d) 0·696      e) 21·007
   f) 300·004     g) 490·255     h) 803·506    i) 267·489

2) The number **1·291** has **1 whole and 291 thousandths**.

   Write in numerals:

   a) 5 wholes and 712 thousandths
   b) 0 wholes and 136 thousandths
   c) 64 wholes and 505 thousandths
   d) 11 wholes and 9 thousandths
   e) 95 wholes and 80 thousandths

| Ones | | | | Decimal Fractions | | |
|---|---|---|---|---|---|---|
| H | T | O | | $\frac{1}{10}$ | $\frac{1}{100}$ | $\frac{1}{1000}$ |
| | | 1 | • | 2 | 9 | 1 |

3) Sebastian Vettel won the 2018 British Grand Prix in one hour, 27 minutes and 29·784 seconds. Lewis Hamilton was second, taking 2·264 seconds longer than Vettel to complete the race.

   Hamilton's result was recorded as +2·264 seconds. This means he finished 2·264 seconds after Vettel.

   Write what these drivers' results mean.

   a) Kimi Raikkonen +3·652 seconds
   b) Valtteri Bottas  +8·883 seconds
   c) Daniel Ricciardo +9·500 seconds
   d) Nico Hülkneberg  + 28·220 seconds

**CHALLENGE!**

At the 2018 European Championships, British athlete Zharnel Hughes won the 100 m sprint in a championship record time of 9·950 seconds. Nuria thinks this means nine seconds and 95 thousandths of a second. Is she correct? Explain.

## 2.8 Representing and describing decimal fractions

**Before we start**

Explain why each model represents 2·03.

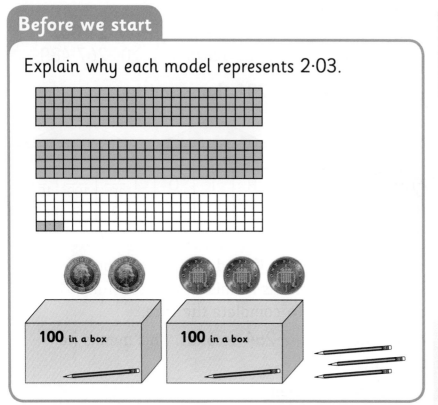

We are learning to build and describe decimal fractions with thousandths.

We can represent decimal fractions with thousandths in different ways.

**Let's learn**

This model represents the decimal fraction 2·138.

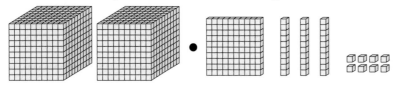

The decimal point helps us work out the value of the base 10 blocks.

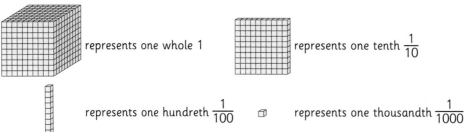

represents one whole 1

represents one tenth $\frac{1}{10}$

represents one hundreth $\frac{1}{100}$

represents one thousandth $\frac{1}{1000}$

## Let's practise

1) Write the decimal fraction represented by the following models in three ways. For example:

2 ones, 1 tenth, 3 hundredths and 8 thousandths = 2·138 = $2\frac{138}{1000}$.

a)

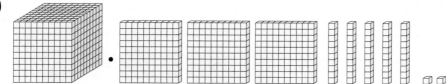

b)

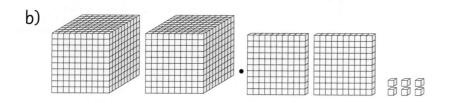

c)

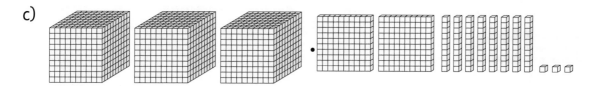

d)

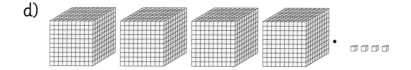

e)

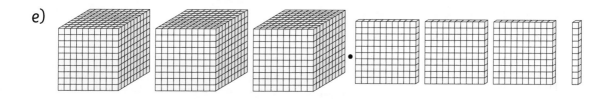

---

## CHALLENGE!

Work with a partner. Build a model of a decimal fraction with three decimal places. Your model should show the value of each digit. Ask another pair to guess the decimal fraction you have created. Can you guess the value of their model?

## 2.9 Zero as a placeholder in decimal fractions

We are learning to use zero correctly in decimal fractions.

### Before we start

Isla thinks it is impossible to rearrange these cards to make more than eight six-digit numbers. Do you agree? Prove it!

0 0 7 0 5 0

We use zero as a placeholder in some decimal fractions.

### Let's learn

The number **7·049** has **zero in the tenths place**. Zero keeps the hundredths and thousandths digits in place. 7·049 = 7 ones, 0 tenths, 4 hundredths and 9 thousandths.

| Ones | | | | Decimal Fractions | | |
|---|---|---|---|---|---|---|
| H | T | O | | $\frac{1}{10}$ | $\frac{1}{100}$ | $\frac{1}{1000}$ |
| | | 7 | • | 0 | 4 | 9 |

The number **1·506** has **zero in the hundredths place**. Zero keeps the tenths and thousandths digits in place. 1·506 = 1 one, 5 tenths, 0 hundredths and 6 thousandths.

| Ones | | | | Decimal Fractions | | |
|---|---|---|---|---|---|---|
| H | T | O | | $\frac{1}{10}$ | $\frac{1}{100}$ | $\frac{1}{1000}$ |
| | | 1 | • | 5 | 0 | 6 |

The number **2·003** has **zero in the tenths and hundredths places**. The zeros keep the thousandths digit in place. 2·003 = 2 ones, 0 tenths, 0 hundredths and 3 thousandths.

| Ones | | | | Decimal Fractions | | |
|---|---|---|---|---|---|---|
| H | T | O | | $\frac{1}{10}$ | $\frac{1}{100}$ | $\frac{1}{1000}$ |
| | | 2 | • | 0 | 0 | 3 |

**Let's practise**

1) Describe the position of the placeholders in these decimal fractions. For example:

38·001 → the placeholders are in the tenths and hundredths places

a) 40·327      b) 80·015      c) 26·009      d) 94·402

e) 106·075      f) 657·009      g) 430·201      h) 500·670

2) This diagram represents 1 whole and 51 hundredths. We can write this as a fraction or as a decimal fraction.

$1\frac{51}{100}$ one and fifty-one hundredths

**1·51** *one point five one*

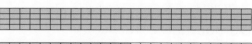

Decide whether each statement is true or false. Rewrite the false statements to make them true.

a) $2·4 = 2\frac{4}{100}$      b) $9·007 = 9\frac{7}{1000}$      c) $12·03 = 12\frac{3}{10}$

d) $10·09 = 10\frac{9}{100}$      e) $6·541 = 6\frac{541}{1000}$      f) $5·055 = 5\frac{55}{1000}$

3) Finlay draws a diagram to prove that $\frac{60}{100}$ is equivalent to $\frac{6}{10}$.

Make a model or draw a diagram to prove that:

a) 0·4 = 0·40          b) 0·5 + 0·09 = 0·59

c) 0·2 + 0·07 = 0·27      d) 0·8 + 0·01 = 0·81

Sixty hundredths equal six tenths so **0·60 and 0·6 mean the same**.

Which zero *isn't* a place holder?

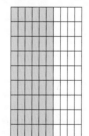

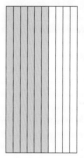

**CHALLENGE!**

Copy and complete each statement by writing = or ≠ in the box. Discuss your answers with a partner and check that you agree.

3·40 ☐ 3·4        0·07 ☐ 0·70        0·29 ☐ 0·290

## 2.10 Partitioning decimal fractions

We are learning to partition decimal fractions with up to three decimal places.

**Before we start**

The children are discussing different ways to partition the number 52071. Who is correct? Explain.

It's 52 thousands and 71 ones.

It's 50000 + 2071.

It's 50000 + 2000 + 0 + 71.

It's 50000 + 2000 + 70 + 1.

Decimal fractions can be partitioned just like whole numbers can.

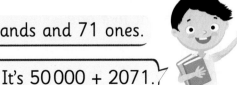

**Let's learn**

Finlay and Amman are investigating decimal fractions using place value arrow cards. They know that:

0·005 means 0 ones, 0 tenths, 0 hundredths and 5 thousandths  `0·0 0 5`

0·03 means 0 ones, 0 tenths and 3 hundredths `0·0 3`

0·8 means 0 ones and 8 tenths `0·8`

9. means 9 ones `9·`

They use the cards to make the decimal fraction 9·835 `9· 8 3 5`

**9·835 = 9 + 0·8 + 0·03 + 0·005**

## Let's practise

1) Write the decimal fraction that can be made using these place value arrow cards.

a) $\boxed{5\cdot}$ $\boxed{0\cdot1}$ $\boxed{0\cdot0\,2}$ $\boxed{0\cdot0\,0\,6}$

b) $\boxed{8\cdot}$ $\boxed{0\cdot3}$ $\boxed{0\cdot0\,4}$ $\boxed{0\cdot0\,0\,9}$

c) $\boxed{1\cdot}$ $\boxed{0\cdot7}$ $\boxed{0\cdot0\,6}$ $\boxed{0\cdot0\,0\,6}$

d) $\boxed{4\cdot}$ $\boxed{0\cdot6}$ $\boxed{0\cdot0\,8}$ $\boxed{0\cdot0\,0\,2}$

e) $\boxed{0\cdot}$ $\boxed{0\cdot5}$ $\boxed{0\cdot0\,7}$ $\boxed{0\cdot0\,0\,1}$

f) $\boxed{0\cdot0}$ $\boxed{0\cdot0\,5}$ $\boxed{0\cdot0\,0\,3}$

Predict the answer then use place value arrow cards to check.

2) Partition these decimal fractions. For example: **6·42 = 6 + 0·4 + 0·02**

a) 3·8      b) 5·5      c) 7·9

d) 2·6      e) 4·52     f) 8·31

g) 1·17     h) 6·29     i) 9·07

j) 2·246    k) 7·403    l) 3·098

m) 6·009

Place value arrow cards can help here too!

3) Write the number represented by each

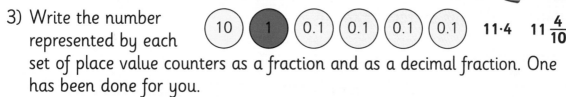

$\bigcirc$ 10 ● 1 $\bigcirc$ 0.1 $\bigcirc$ 0.1 $\bigcirc$ 0.1 $\bigcirc$ 0.1    **11·4**    **11$\frac{4}{10}$**

set of place value counters as a fraction and as a decimal fraction. One has been done for you.

a)

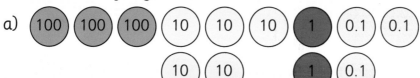

b)

c)

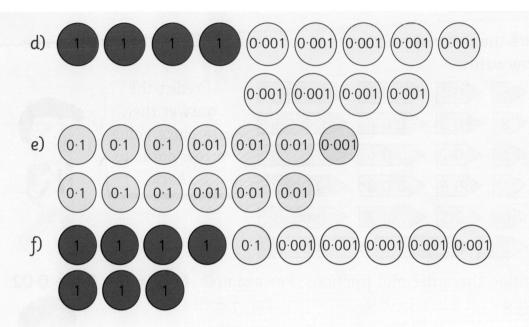

d)

e)

f)

★ **CHALLENGE!**

Mr Smith has asked the children to make 0·012 using place value counters. Who is correct? Explain.

Amman: (0·01) (0·001) (0·001)

Finlay: (0·1) (0·001) (0·001)

Isla: (0.01) (0.01) (0.01) (0.01) (0.01) (0.01)
(0.01) (0.01) (0.01) (0.01) (0.01) (0.01)

Nuria: (0·001) (0·001) (0·001) (0·001) (0·001) (0·001)
(0·001) (0·001) (0·001) (0·001) (0·001) (0·001)

## 2.11 Comparing and ordering decimal fractions

We are learning to compare decimal fractions and put them in order.

**Before we start**

Nuria thinks she has ordered these numbers from largest to smallest but she has made a few mistakes. Explain to Nuria what she has done wrong.

| 61·4 | 406 | 6·4 | 46 | 1·6 | 0·4 |

Write what Nuria should have written.

We can compare and order decimal fractions by thinking about the value of each digit.

**Let's learn**

To compare decimal fractions we begin by looking at the whole number. If two decimal fractions share the same whole number we look at the digits after the decimal point to say which is larger.

0·187 and 0·718 each have three digits after the decimal point.

0·718 > 0·187 because 718 thousandths is more than 187 thousandths.

It is important to think carefully about the values of the digits. Let's compare 0·13 and 0·3.

0·13 is 0 ones and 13 hundredths.

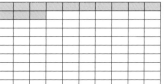 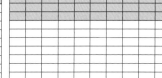

0·3 is 0 ones and 3 tenths = 30 hundredths.

So **0·3 > 0·13** and **0·13 < 0·3**.

## Let's practise

1) Write these decimal fractions in order from smallest to largest.

    a) 4·109       9·104       1·094       4·091       1·049

    b) 3·629       6·926       9·926       3·926       9·326

    c) 5·552       5·255       2·515       1·255       5·125

    d) 0·689       0·869       6·986       8·009       6·989

2) Isla is thinking of a decimal fraction. She gives Amman three clues and challenges him to guess her number.

```
|.....|.....|.....|.....|.....|.....|.....|.....|.....|.....|
2    2·1   2·2   2·3   2·4   2·5   2·6   2·7   2·8   2·9    3
```

    a) What could Isla's number be? List all the possibilities.

    b) Choose one of the possible answers and write it down. Write three clues and challenge a partner to guess which number you have picked.

My number has two decimal places.

It sits between 2 and 3 on the number line.

It is more than 2·4 but less than 2·5.

3) Write down whether the following statements are true or false:

Drawing or imagining a number line may help.

    a) 13·0 > 13        b) 1·89 < 1·8

    c) 5·18 = 5·81     d) 6·4 < 6·40

    e) 3·09 = 3·9      f) 24 < 20·4

    g) 4·01 < 4·1      h) 5 = 5·000

4) Look again at the false statements in question 3. Change the symbol in each example to make the statement true.

5) Write in order from largest to smallest.

    a) 26·5, 2·65, 2·5, 25, 26·05, 2·06

    b) 1·78, 17·8, 178, 7·08, 7·81, 1·7

    c) 34, 9, 3·49, 0·94, 4·09, 43

6) 1 ten is 10 times smaller than
1 hundred.
1 one is 10 times smaller than
1 ten.
1 tenth is 10 times smaller
than 1 one.
1 hundredth is 10 times smaller
1 tenth.
1 thousandth is 10 times smaller
than 1 hundredth.
**10 thousandths = 1 hundredth**
**10 hundredths = 1 tenth**      **10 tenths = 1**

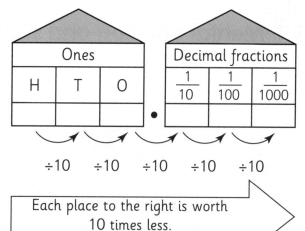

Copy and complete the following statements. One has been done for you.

a)  8 hundredths and 6 thousandths = [ 86 ]  thousandths = [ 0·086 ]

b)  7 hundredths and 2 thousandths = [   ]  thousandths = [   ]

c)  3 tenths and 6 hundredths and 0 thousandths = 36 [   ]  = [   ]

d)  500 thousandths = [   ]  hundredths = [   ]  tenths = 0·5

7) Arrange these numbers in order from smallest to largest.

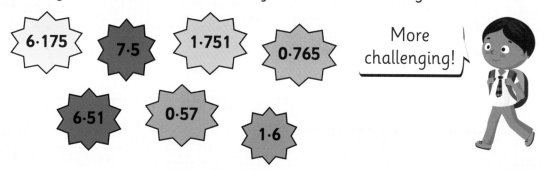

6·175   7·5   1·751   0·765   More
challenging!

6·51   0·57   1·6

⭐ **CHALLENGE!** ..........................................................

If you were to count up in tenths from 0·1 to 19·7, how many tenths
would you count?

## 3.1 Mental addition and subtraction

We are learning to choose the most efficient strategy to add and subtract whole numbers mentally.

### Before we start

Mrs Stewart challenges Finlay to work out the answers to these calculations mentally. Explain what Finlay should do then find each answer using your chosen strategy.

a) 796 + 419　　b) 5062 – 37　　c) 8000 – ☐ = 988

Discuss our strategies with a partner. We jotted down the numbers in bold to help us keep track of our thinking.

### Let's learn

**10 351 + 10 351**

Double 10 000 is **20 000**
Double 350 is **700**
Double 1 is **2**

My answer is 20 702.

**3278 + 15 993**

16 000 plus 3278 is **19 278**
I've added 7 too many so I need to compensate.
19 278 – **7** equals 19 271

My answer is 19 271.

**746 061 – 997**

746 061 – 997 is the same as **746 064 – 1000**

My answer is 745 064.

**14 000 – 13 826**

13 826 add **4** makes 13 830
Add **70** takes me to 13 900
Add **100** more to make 14 000

My answer is 174.

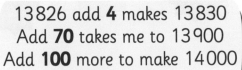

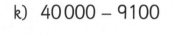

## Let's practise

1) Calculate using a mental strategy of your choice. Explain to a partner how you worked each answer out.

   a) 16 358 + 999     b) 81 028 + 595     c) 201 + 77 427
   d) 11 039 + 1601   e) 3008 + 14 217   f) 5998 + 26 000
   g) 27 500 – 491    h) 19 205 – 998
   i) 33 162 – 137     j) 56 004 – 1088  Don't forget to
   k) 40 000 – 9100    l) 60 845 – 9003 explain your jottings!

2) Place value and number facts are very helpful when calculating mentally. Let's calculate **175 431 + 20 030**.

We need to add 20 thousands, 0 hundreds, 3 tens and 0 ones.
We know that adding zero to a number does not change its value, therefore the only digits that will change are the tens of thousands digit and the tens digit.

175 4**3**1 + **2**0 000 = 1**9**5 4**3**1     195 4**3**1 + **3**0 = 195 4**6**1

Use place value and number facts to calculate:

   a) 305 091 + 80 000      b) 153 069 + 104 000
   c) 872 816 + 13 003      d) 163 595 + 300 200
   e) 402 063 + 86 225      f) 104 300 + 683 014
   g) 386 562 – 20 000      h) 948 601 – 506 000
   i) 475 726 – 15 001      j) 591 504 – 140 500
   k) 649 270 – 27 030      l) 830 187 – 210 004

## CHALLENGE!

Is Finlay correct? Explain. It may help to draw place value houses.

Two million, seven hundred thousand and eighty-five **minus** one million, four hundred thousand and seventy equals one million, three hundred thousand and fifteen.

## 3.2 Adding and subtracting a string of numbers

We are learning to add and subtract a string of numbers by making groups.

### Before we start

Which **three** numbers total 6040?

1860

4240

3413

1003

2003

1087

3177

540

Grouping numbers together can make it easier to add or subtract them mentally.

### Let's learn

Let's add **48 + 170 + 530 + 4000 + 152 + 8000**

Look for pairs or groups that total a multiple of 10, 100 or 1000.
12 000 + 700 + 200 = 12 900.

48 + 170 + 530 + 4000 + 152 + 8000

**The answer is 12 900.**

Let's subtract **15 250 – 215 – 35 – 190 – 410**

Look for pairs of numbers that total a multiple of 10, 100 or 1000 and subtract them together.

250        600

15 250 – 215 – 35 – 190 – 410

15 250 – 250 = 15 000 – 600 = 14 400. **The answer is 14 400**.

## Let's practise

1) Add these strings of numbers.

a) 237 + 22 + 43 + 1000 + 9000 + 558
b) 70 + 90 + 8000 + 12000 + 410 + 60
c) 5300 + 499 + 7000 + 1700 + 13000
d) 34 + 1300 + 90 + 610 + 156 + 1700
e) 18000 + 5000 + 2000 + 190 + 5000
f) 33000 + 350 + 350 + 27000 + 1019
g) 4700 + 710 + 590 + 1300 + 248 + 32

Write the numbers down and cross them off as you use them. Jottings help you keep track of your thinking.

2) Subtract each set of numbers from the starting number in bold.

a) Start with **14095**. Subtract 260, 526 and 214 from this number.
b) Start with **23726**. Subtract 145, 255 and 900 from this number.
c) Start with **20000**. Subtract 4000, 2000 and 8000 from this number.
d) Start with **17100**. Subtract 1017, 340 and 683 from this number.
e) Start with **48080**. Subtract 62, 18, 4100 and 900 from this number.

3) Fill in the missing digits. Write the completed number sentences.

a) 1275 + 2✳✳✳ + 803 = 4503
b) 9✳✳✳ − 360 − 240 = 9073

CHALLENGE!

Using each digit once only, write two four-digit numbers that total 10000. How many different ways can you find?

| 1 | 2 | 2 | 5 | 5 | 7 | 7 | 8 |

# Number – addition and subtraction

## 3.3 Using place value partitioning to add and subtract

We are learning to add and subtract six-digit numbers by partitioning them into thousands, hundreds, tens and ones.

**Before we start**

How many of each counter is needed to make:

a) forty-eight thousand, two hundred and forty

b) sixty thousand, five hundred and five

c) twenty-three thousand and seventeen

d) five hundred thousand and one

Partitioning helps us with addition and subtraction.

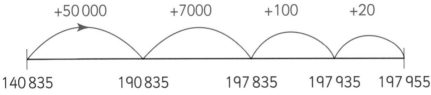

**Let's learn**

Nuria uses an empty number line to work out **57 120 + 140 835**.

She starts with 140 835 and partitions 57 120 into 50 000 + 7000 + 100 + 20.

```
        +50 000        +7000        +100         +20
      ⌒            ⌒           ⌒            ⌒
  |_____|_____|_____|_____|
140 835        190 835      197 835    197 935  197 955
```

**57 120 + 140 835 = 197 955**

Finlay uses the column method to
work out **848 376 − 613 228**.

**848 376 − 613 228 = 235 148**

$$
\begin{array}{r}
848\,376 \\
-\ 613\,228 \\
\hline
200\,000 \\
30\,000 \\
5\,000 \\
100 \\
50 \\
-\quad 2 \\
\hline
235\,148
\end{array}
$$

## Let's practise

1) Use Nuria's method to calculate:

a) 102 462 + 57 516
b) 84 260 + 610 539
c) 273 055 + 133 333
d) 93 697 – 42 325
e) 177 616 – 41 209
f) 686 581 – 632 150

2) Use Finlay's method to calculate:

a) 62 561 + 26 746
b) 149 275 + 30 484
c) 350 219 + 196 523
d) 67 843 – 27 457
e) 258 746 – 141 537
f) 967 979 – 532 180

Don't forget to subtract the negative numbers!

3) Choose Nuria or Finlay's method to work out:

a) 222 585 + 608 974
b) 533 574 + 365 289
c) 208 182 + 394 236
d) 875 491 – 160 423
e) 659 343 – 121 268
f) 194 635 – 178 135

### CHALLENGE!

One thousand thousands = 1 million = 1 000 000. Match pairs of cards to make one million.

| | | | |
|---|---|---|---|
| 829 900 | 400 000 | 215 000 | 600 000 |
| 421 060 | 160 000 | 578 940 | 785 000 |

## 3.4 Adding whole numbers using standard algorithms

We are learning to add five-digit and six-digit numbers using a standard written method.

### Before we start

Identify the errors in these additions, then correct them.

a)
```
   1 1
  4 5 8 9
+ 1 3 4 6
---------
  5 9 2 5
```

b)
```
 1 1   1
    7 7 5 7
+   4 3 2 5
-----------
    2 0 8 2
```

c)
```
  1 1 1
  2 6 7 4
+ 3 7 6 6
---------
    6 4 4
```

Algorithms can help us with addition calculations that are too tricky to work out mentally.

### Let's learn

There are 16 ones = **1 ten** and 6 ones.
Write 6 in the ones column and carry **1 ten**.

There are 12 tens = **1 hundred** and 2 tens.
Write 2 in the tens column and carry **1 hundred**.

There are 24 hundreds = **2 thousands** and 4 hundreds.
Write 4 in the hundreds column and carry **2 thousands**.

There are 16 thousands = **1 ten thousand** and 6 thousands.
Write 6 in the thousands column and carry **1 ten thousand**.

There are 9 ten thousands.

```
  1 2   1 1
  4 8   9 1 7
  3 2   9 4 5
+ 1 4   5 6 4
-------------
  9 6   4 2 6
```

**Let's practise**

1) Set each addition down as an algorithm. Use Amman's method to work out the answers.

   a) 31 562 + 82 795
   b) 72 843 + 53 977
   c) 23 756 + 199 384
   d) 706 319 + 55 927
   e) 496 276 + 186 528
   f) 534 435 + 331 829
   g) 273 858 + 608 767
   h) 444 553 + 764 764

2) Calculate the answers to these questions using the standard written algorithm for addition.

   a) 21 480 + 45 918 + 17 158
   b) 37 541 + 59 652 + 64 271
   c) 42 759 + 12 921 + 17 316
   d) 81 778 + 11 363 + 72 867
   e) 115 906 + 31 476 + 3286
   f) 4895 + 196 543 + 9384

3) a) Using each of the digits below only once each time, write an addition algorithm where you need to:
      i)   carry a thousand into the ten thousands column
      ii)  carry a ten into the hundreds column
      iii) carry over a ten, a hundred and a thousand
   b) Ask a partner to find the answers to your additions.
   c) Work out the answers to the additions your partner has written.

   2   6   7   3   9   5   1   4   0   8

---

⭐ **CHALLENGE!**

Copy and complete these algorithms by filling in the missing digits.

```
    ❋11 ❋29              2❋6 ❋13
  + 2❋5 07❋            + 10❋ 828
  ─────────            ─────────
    796 505              312 24❋
```

## 3.5 Subtracting whole numbers using standard algorithms

> We are learning to subtract five-digit and six-digit numbers using a standard written method.

**Before we start**

Amman has made some mistakes with these subtractions. Write down what Amman should have written.

a)
```
      3 1 2 1
    4 0 3 2
  – 2 3 5 6
  ─────────
    1 7 3 6
```

b)
```
      2 1
    8 3 0 7
  – 1 1 4 8
  ─────────
    7 1 6 1
```

c)
```
    6 11 1 1
    5 2 0 1
  – 3 8 6 3
  ─────────
    3 3 4 8
```

> An algorithm can help us with subtraction calculations that are too tricky to work out mentally.

**Let's learn**

9 ones – 7 ones = 2 ones. Write 2 in the ones column.

I can't take 4 tens from 0 tens.

**Exchange a hundred for 10 tens and add them to 0 tens.**

10 tens – 4 tens = 6 tens.

I can't take 5 hundreds from 1 hundred.

**Exchange a thousand for 10 hundreds and add them to 1 hundred.**

11 hundreds – 5 hundreds = 6 hundreds.

3 thousands – 1 thousand = 2 thousands.

6 ten thousands – 5 ten thousands = 1 ten thousand.

```
          3  11
    6 4  2 0 9
  – 5 1  5 4 7
  ───────────
    1 2  6 6 2
```

## Let's practise

1) Set each subtraction down as an algorithm. Use Nuria's method to work out the answers.

   a) 27 706 – 14 921
   b) 260 327 – 54 112
   c) 348 863 – 33 602
   d) 456 312 – 235 821
   e) 195 237 – 36 152
   f) 776 567 – 130 973
   g) 525 020 – 121 508
   h) 856 324 – 513 361

2) Now try these more challenging questions. Check your answers by adding.

   a) 65 316 – 35 822
   b) 81 032 – 25 367
   c) 156 309 – 72 538
   d) 450 390 – 72 165
   e) 573 701 – 281 446
   f) 916 803 – 308 588
   g) 220 086 – 134 391
   h) 701 203 – 467 315

3) a) Using each digit only once in each question, write a subtraction algorithm where you need to:

   i)   exchange 1 thousand for 10 hundreds
   ii)  exchange 1 ten thousand for 10 one thousands
   iii) exchange 1 ten thousand for 10 thousands **and** 1 ten for 10 ones.

   | 2 | 6 | 9 | 3 | 0 | 8 | 1 | 4 | 7 | 5 |
   |---|---|---|---|---|---|---|---|---|---|

   b) Ask a partner to find the answers to your subtractions.
   c) Work out the answers to the subtractions your partner has written.

## CHALLENGE!

Copy and complete these algorithms by filling in the missing digits.

```
  ✳1 7 6 5✳          6✳0 ✳3 1
– 2✳9 5 8 3        – ✳7 5 2 8 5
 ─────────          ─────────
  6 0 8 0✳5          4 4 5 1✳✳✳
```

## 3.6 Mental and written calculation strategies

We are learning to choose the most efficient strategy to solve an addition or subtraction problem.

**Before we start**

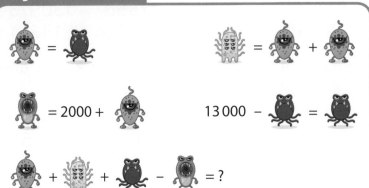

There are many strategies that we can use to help us solve number problems.

**Let's learn**

The correct choice of strategy helps us to solve number problems efficiently.

Isla thinks the most efficient way to work out **130 000 + 40 000** is to use a standard algorithm. Do you agree?

Finlay thinks the most efficient way to work out **57 500 – 499** is to use 'round and adjust'. Do you agree?

Amman thinks the most efficient way to work out **4886 + 7867** is to use a standard algorithm. Do you agree?

**Let's practise**

1) Calculate the following. Explain how you worked each answer out.

a) 250 000 + 250 000

b) 152 000 + 341 998

c) 654 637 + 175 877

d) 180 000 + 9999

e) 497 655 + 456 709

f) 4536 + 7348 + 1152

2) Find the missing number on each brick by calculating the difference between the two numbers directly below it. Think carefully about your choice of strategy.

| 25 763 | 18 448 | 13 529 | 9 938 | 6 538 |
|---|---|---|---|---|

⭐ **CHALLENGE!**

With a partner, play **Three in a Row**.

- Take turns to choose an addition or subtraction from the list below and find the answer using the most efficient strategy.
- Cover the answer on the grid with a counter of your colour.
- The first player to get three counters in a row, horizontally, vertically or diagonally, is the winner.

| | | | |
|---|---|---|---|
| 11 340 | 8 239 | 13 040 | 7 211 |
| 11 074 | 6 276 | 9 810 | 3 112 |
| 8 374 | 11 974 | 10 831 | 9 675 |
| 17 632 | 5 094 | 11 264 | 6 060 |

7774 + 4200       11 931 – 1100

3567 + 4672       19 750 – 8410

2879 + 8385       15 587 – 12 475

2650 + 7025       24 831 – 17 620

4187 + 4187       30 094 – 25 000

6039 + 5035       18 280 – 12 220

4767 + 8273       10 000 – 190

13 829 + 3803     12 276 – 6000

## 3.7 Representing word problems

We are learning to represent and solve the same word problem in different ways.

**Before we start**

Isla is stuck on this word problem. Explain what she should do then solve it.

*A football stadium can hold 24 000 people. If there are 9486 fans in the away stand and 13 566 fans in the home stand, how many empty seats are there?*

Representing the same problem in different ways can help to deepen our understanding.

**Let's learn**

Amman draws a Think Board to help him represent and solve a word problem.

| Word problem | Bar model |
|---|---|
| *The V&A museum in Dundee aims to attract 500 000 visitors during its first year of opening. So far, there have been 126 100 visitors. How many more people need to visit for the museum to reach its target?* | 500 000 / 126 100 / ? |

**Answer**

**373 900**

**Empty number line**

+900  +3000  +70 000  +300 000

126 100  127 000  130 000  200 000      500 000

**Calculation(s)**

−100 000 = 400 000

−20 000 = 380 000

−6000 = 374 000

−100 = 373 900

### Let's practise

Draw Think Boards like Amman's to represent and solve these word problems.

1) 247 139 people visited the Cairngorms last year and some people visited Ben Nevis. If 137 139 more visited the Cairngorms than Ben Nevis, how many people visited Ben Nevis?

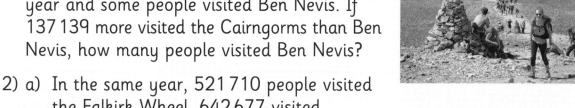

2) a) In the same year, 521 710 people visited the Falkirk Wheel, 642 677 visited Edinburgh Zoo and 567 259 visited Stirling Castle. How many visitors is this?

   b) Stirling Castle received 187 107 more visitors than Urquhart Castle. How many people visited Urquhart Castle?

3) Edinburgh Castle received 2 063 709 visitors. Is this more or less than one million? How many more or less?

4) The National Museum of Scotland, Edinburgh, received 2 165 601 visitors. The Riverside Museum, Glasgow, received 1 355 359 visitors. How many more people visited the Edinburgh museum?

### CHALLENGE!

Write your own word problems with at least one six-digit number in them. Challenge a partner to draw Think Boards and solve them.

## 3.8 Solving multi-step word problems

We are learning to choose the most efficient method to solve a word problem.

### Before we start

Solve these word problems. Explain how you worked them out.

a) How many fewer people live in the least populated city than in the most populated city?

b) How many more people live in Edinburgh than in Aberdeen and Dundee combined?

| City | Population in 2018 |
|------|-------------------|
| Aberdeen | 207 932 |
| Dundee | 148 270 |
| Edinburgh | 482 005 |
| Glasgow | 598 830 |
| Perth | 43 450 |

When solving a challenging problem, it helps if we can break it down into steps.

### Let's learn

Isla draws a bar model to help her think about this word problem.

*47 348 votes are cast in a TV reality show. Contestant Number 1 receives 13 721 votes. Contestant Number 2 receives 18 167 votes.*

*Contestant Number 3 and Contestant Number 4 receive exactly the same number of votes. How many votes did they each receive?*

| 47 348 | | | |
|--------|--------|-----|-----|
| 13 721 | 18 167 | ? | ? |

She does the following calculations:

$$\begin{array}{r} \overset{1}{1}\,3\ 7\ 2\ 1 \\ +\,1\,8\ 1\ 6\ 7 \\ \hline 3\,1\ 8\ 8\ 8 \end{array} \qquad \begin{array}{r} 4\ \overset{6}{\cancel{7}}\ \overset{12}{\cancel{3}}\ \overset{1}{4}\ 8 \\ -\,3\ 1\ 8\ 8\ 8 \\ \hline 1\,5\ 4\ 6\ 0 \end{array} \qquad \frac{1}{2} \text{ of } 15\,460 = 7730$$

**Contestants 3 and 4 get 7730 votes each.**

**Let's practise**

1) a) Nuria's family buy a house for £175 200, which is £6800 less than the original asking price. What was the original asking price?

   b) The builder spent £125 066 on materials and £39 500 on labour. If he sold the house for £175 200, what was his profit?

2) The children are playing a computer game. Amman scores 72 156 points. Isla scores 77 849 points. Finlay scores 1294 points more than Amman. Nuria scores 573 points less than Isla. What was the children's total score?

3) A vets' practice treats a total of 723 patients in one week. $\frac{1}{3}$ are cats, 387 are dogs and the rest are small creatures such as hamsters, gerbils and mice. How many small creatures were treated?

4) Finlay and his family are flying to New Zealand to visit his Auntie Jacqui. The total flight distance from Aberdeen to New Zealand is 19 951 km. Aberdeen to London is 640 km. By the time they reach Singapore they still have 8462 km to go. How far is it from London to Singapore?

**CHALLENGE!**

Write your answers in numerals and in words.

1) How much further from the Sun is
   a) Venus than Mercury?        b) Mars than Mercury?

2) How much closer to the Sun is Earth than Jupiter?

| Planet | Distance from the sun in kilometres (km) |
|---|---|
| Mercury | 58 000 000 |
| Venus | 108 000 000 |
| Earth | 149 600 000 |
| Mars | 228 000 000 |
| Jupiter | 778 500 000 |

## 3.9 Adding whole numbers and decimal fractions

We are learning to add whole numbers and decimal fractions with two decimal places, mentally.

**Before we start**

Finlay thinks the answer to 37 + 4·5 is 82. Explain what Finlay has done wrong then write the correct answer.

Understanding the value of each digit helps us to add whole numbers and decimal fractions.

**Let's learn**

To calculate **6·42 + 38** Amman **rounds to the nearest 10**.

He knows that 6.42 means 6 ones and 42 hundredths. He rounds 38 up to 40 and **compensates** by taking two away from 6·42. His calculation now says **4·42 + 40**.

| **6·42 + 38 is the same as 4·42 + 40.** | **The answer is 44·42.** |

Isla is working out **3.97 + 5.68**.

She **rounds** 3·97 up **to the nearest whole number** by adding three hundredths **then compensates** for this by subtracting three hundredths from 5·68. Her calculation now says **4 + 5·65**.

| **3·97 + 5·68 is the same as 4 + 5·65.** | **The answer is 9·65.** |

Nuria is working out the answer to **16·52 + 8·14**.

She **partitions** each decimal fraction into whole numbers and hundredths.

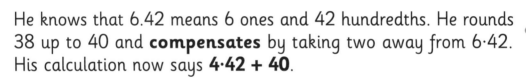

| **16·52 + 8·14 = 16 + 8 + 0·52 + 0·14 = 24 + 0·66 = 24·66** |

## Let's practise

1) Use Amman's strategy to calculate:

   a) 59 + 6·89      b)  7·71 + 39      c)  28 + 3·02
   d) 89 + 5·55      e)  2·64 + 58      f)  1·92 + 99
   g) 69 + 12·31     h)  398 + 14·13    i)  61·85 + 499
   j)  32·22 + 549   k)  239 + 27·91    l)  31·44 + 919

2) Use Isla's strategy to calculate:

   a) 5·99 + 2·08    b)  7·17 + 2·98    c)  4·97 + 9·82
   d) 2·15 + 3·99    e)  6·54 + 6·97    f)  8·98 + 9·77
   g) 1·96 + 6·12    h)  9·96 + 7·08    i)  7·59 + 6·99
   j)  4·99 + 8·99   k)  3·98 + 3·98    l)  2·96 + 5·97

3) Use Nuria's strategy to calculate:

   a) 14·27 + 9·52   b)  26·35 + 5·41   c)  82·18 + 5·39
   d) 78·43 + 17·44  e)  38·38 + 22·51  f)  41·09 + 16·76

4) Calculate using a mental strategy of your choice:

   a) 23·58 + 4·16   b)  2·97 + 11·66   c)  49·9 + 6·08
   d) 17·35 + 299    e)  68 + 112·2     f)  5·99 + 22·98

### CHALLENGE!

a) Match pairs of cards to make the decimal fraction 2670·88.

| 1000·45 | 1670·43 | 1330·87 | 70·88 |
|---------|---------|---------|-------|
| 0·88 | 2600 | 1340·01 | 2670 |

b) Write another **four pairs** of decimal fractions that total 2670·88.

## 3.10 Adding decimal fractions using standard written algorithms

We are learning to add decimal fractions using a standard written method.

### Before we start

Two decimal fractions add up to 1. What could they be? Give three different examples.

Algorithms can help us add decimal fractions that are too tricky to add mentally.

### Let's learn

Finlay is adding 137·86 and 682·65. He thinks carefully about each digit's place value and writes the numbers down in columns. He writes the whole number part to the left of the decimal point and the fraction part to its right.

11 hundredths = 1 tenth and 1 hundredth.
Write 1 in the hundredths column and carry 1 tenth.
15 tenths = 1 one and 5 tenths.
Write 5 in the tenths column and carry 1 one.
10 ones = 1 ten and 0 ones.
Write 0 in the ones column and carry 1 ten.
12 tens = 1 hundred and 2 tens.
Write 2 in the tens column and carry 1 hundred.
1 hundred + 1 hundred + 6 hundreds = 8 hundreds.

```
  1 1 1   1
  1 3 7 · 8 6
+ 6 8 2 · 6 5
-----------
  8 2 0 · 5 1
```

**Let's practise**

1) Use Finlay's method to find the answers to these additions:
   a) 28·62 + 13·75
   b) 66·29 + 14·66
   c) 32·47 + 53·68
   d) 129·92 + 58·69
   e) 330·16 + 88·09
   f) 259·28 + 11·79
   g) 29·74 + 172·38
   h) 19·65 + 470·56
   i) 16·87 + 378·95
   j) 551·16 + 356·27
   k) 896·28 + 421·77
   l) 543·52 + 129·88

2) Calculate the answers to these questions using the standard algorithm for addition:
   a) 6·78 + 15·37 + 0·54
   b) 10·66 + 3·77 + 9·88
   c) 18·03 + 16·76 + 29·88
   d) 49·32 + 85·67 + 77·81

3) Using each digit card below only once each time, write an algorithm with two decimal fractions where:
   a)  i) the answer has the digit '5' in the hundredths column
       ii) the answer is > 100
       iii) you only need to carry one tenth over from the hundredths column
       iv) you need to carry three times
   b) Is it possible to write an algorithm that gives an answer < 50? Explain.

   $\boxed{3}\ \boxed{6}\ \boxed{7}\ \boxed{8}\ \boxed{9}\ \boxed{1}\ \boxed{4}\ \boxed{5}$

⭐ **CHALLENGE!**

Copy and complete.

```
  1 4 ✳ · 2 8          5 2 9 · 8 7          4 ✳ 2 · 8 5
+ ✳ 0 4 · ✳ 3        + 3 8 5 · ✳ 6        + 4 2 ✳ · 3 5
  ─────────            ─────────            ─────────
  3 5 0 · 8 1          ✳ 1 ✳ · 0 3          8 2 3 · 2 ✳
```

All of the **answers** have at least one zero in them, but which of the zeros are needed and which are not? Explain your thinking.

## 3.11 Subtracting decimal fractions

We are learning to mentally subtract decimal fractions with two decimal places.

**Before we start**

Calculate mentally. Check each answer by adding.
a) 40·6 – 8·3        b) 68·5 – 20·1
c) 75·8 – 34·8        d) 54·1 – 24

Understanding the value of each digit helps us to subtract decimal fractions mentally.

**Let's learn**

Nuria uses place value partitioning to calculate 9·49 – 3·16.

Nuria's jottings

$6 + \frac{33}{100}$

9 ones subtract 3 ones leaves 6 ones.
I will jot down 6 to remind me how many I have left.
49 hundredths subtract 16 hundredths leaves 33 hundredths. I will jot that down, too.
My answer is $6\frac{33}{100}$ = 6·33.

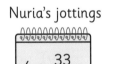

Isla imagines an Empty Number Line and counts on to calculate 15 – 9·74.

Isla jottings

$\frac{6}{100}$   $\frac{20}{100}$   5

I will count up to the next whole number (10).
74 hundredths + 6 hundredths is eighty hundredths. Jot down $\frac{6}{100}$.
80 hundredths + 20 hundredths is 100 hundredths = 1 whole. Jot down $\frac{20}{100}$.
10 + 5 = 15. Jot down **5**.

**Let's practise**

1) Use Nuria's method to calculate:

    a) 8·72 – 4·15         b) 9·64 – 3·59         c) 8·15 – 3·07

    d) 20·66 – 9·42       e) 18·65 – 9·23       f) 11·77 – 7·31

    g) 25·59 – 19·44     h) 16·53 – 10·29    i) 35·84 – 35·03

    j) 165·48 – 5·22     k) 180·75 – 7·11     l) 244·93 – 44·03

2) Use Isla's method to calculate:

    a) 20 – 7·46           b) 50 – 9·28          c) 90 – 21·96

    d) 64 – 22·17         e) 78 – 27·32        f) 35 – 7·51

    g) 100 – 74·23       h) 400 – 55·55      i) 700 – 86·03

    j) 900 – 303·98     k) 1000 – 92·29    l) 2000 – 999·41

3) a) Use the numbers on the starbursts to write six different subtractions. One has been done for you: 4·29 – 1·03.

    b) Now calculate the answers to the subtractions you have made. Explain your strategy for each calculation.

**CHALLENGE!**

A **palindrome** is a number that reads the same forwards and backwards, for example 65·56.

It is the difference between 30 and 8·88.

a) What palindrome is Amman describing?

b) Write clues like Amman's for these palindromes. Ask a partner to check them.

    47·74         39·93        80·08

## 3.12 Subtracting decimal fractions using standard algorithms

We are learning to subtract decimal fractions using a standard written method.

**Before we start**

Copy and complete these calculations.
Check each answer by adding.

```
  9 1 3  8 2 0        6 2 1  0 8 7
– 1 4 5  2 3 6      – 4 1 1  5 9 1
  □□□  □□□            □□□  □□□
```

Algorithms can help us with subtraction calculations that are too tricky to work out mentally.

**Let's learn**

Finlay is subtracting 17·85 from 34·72. He thinks carefully about each digit's place value and writes the numbers down in columns. He writes the whole number part to the left of the decimal point and the fraction part to its right.

I can't take 5 hundredths from 2 hundredths.

Exchange a tenth for 10 hundredths.

12 hundredths – 5 hundredths = 7 hundredths.

I can't take 8 tenths from 6 tenths. Exchange a one for 10 tenths. 16 tenths – 8 tenths = 8 tenths.

I can't take 7 ones from 3 ones. Exchange a ten for 10 ones. 13 ones – 7 ones = 6 ones.

2 tens – 1 ten = 1 ten.

```
   2 13 16 1
   3̶ 4̶ · 7̶ 2
 – 1 7 · 8 5
 ─────────────
   1 6 · 8 7
```

**Let's practise**

1) Use Finlay's method to calculate the following. Check your answers by adding.

a)  28·11 – 5·48    b)  19·24 – 2·68    c)  67·15 – 2·44
d)  16·49 – 9·52    e)  25·81 – 7·52    f)  33·33 – 8·88
g)  29·13 – 23·51   h)  38·05 – 35·46   i)  50·41 – 16·87
j)  64·20 – 35·42   k)  71·46 – 69·77   l)  46·14 – 45·37
m) 93·28 – 78·58    n)  80·15 – 39·96   o)  67·75 – 17·69

2) a)  Compare your answers to question 1 with a partner.

   b)  Look carefully at all answers with zeros in them. Are all of the zeros needed? Explain.

   c)  List the answers that have zero as a placeholder.

3) Look again at questions 1(n) and 1(o). Is there a more efficient way to calculate the answers to these questions? How else could you work them out?

**CHALLENGE!**

a)  Match five pairs of numbers, each with a difference of 13·64.

**49·11**    **62·75**

**72·22**    **17·46**

**31·1**    **16·82**

**21**    **7·36**

**30·46**    **58·58**

b) Write two decimal fractions of your own that have a difference of 13·64.

# 3 Number – addition and subtraction

## 3.13 Adding and subtracting decimal fractions

We are learning to choose the most efficient way to add and subtract decimal fractions with one and two decimal places.

### Before we start

Isla thinks the most efficient way to solve both these number problems is to use standard algorithms. Nuria thinks 'round and adjust' is a more efficient strategy.

$$3·25 + 16·99 \qquad 73·29 – 7·42$$

Do you agree with either of the children? Work out each answer using the most efficient strategy each time.

When choosing a strategy, we need to think carefully about the place value of the digits.

### Let's learn

Isla adds this string of numbers using a standard algorithm:
**2·57 + 0·8 + 3·96 + 12·7**

She thinks about the value of each digit, writes them in the correct columns, then solves the problem.

Nuria chooses to do this subtraction mentally: **14·81 – 0·5**

```
   13  1
     2 · 5 7
     0 · 8 0
     3 · 9 6
+  1 2 · 7 0
   ─────────
   2 0 · 0 3
```

5 tenths = 50 hundredths

81 hundredths – 50 hundredths = 31 hundredths

The answer is 14.31.

How could Isla's strategy be used for this subtraction: 126·2 – 78·87?

How could Nuria's strategy be used for this addition: 0·7 + 134·15?

**Let's practise**

1) Work out the answers to these additions using the most efficient method:

   a) 12·34 + 4·1
   b) 18·52 + 9·5
   c) 7·3 + 44·05
   d) 22·6 + 31·18
   e) 77·4 + 24·75
   f) 22·99 + 5·8
   g) 153·1 + 6·68
   h) 500·04 + 0·73
   i) 604·32 + 10·89
   j) 8·63 + 6·8 + 4·61
   k) 6·7 + 7·8 + 10·2 + 15·34

2) Work out the answers to these subtractions using the most efficient method:

   a) 4·1 – 2·65
   b) 8·3 – 7·72
   c) 6·1 – 0·03
   d) 26·38 – 14·1
   e) 14·6 – 1·99
   f) 50·6 – 22·01
   g) 75·66 – 20·3
   h) 61·4 – 29·57
   i) 44·09 – 3·3
   j) 200 – 80·2
   k) 160·9 – 15·55
   l) 802·36 – 58·6

3) Write down what you must add or subtract from each input number to reach the output number. One has been done for you.

   **INPUT** **OUTPUT** **INPUT** **OUTPUT**

   a) 83·12 [ –1 ] 82·12
   b) 41·36 [ ] 41·86

   c) 62·43 [ ] 62·49
   d) 72·01 [ ] 72·3

   e) 57·29 [ ] 57·99
   f) 12·43 [ ] 14·53

   g) 57·6 [ ] 57·25
   h) 35·15 [ ] 34·05

   i) 17·88 [ ] 11·82
   j) 179·01 [ ] 176·71

**CHALLENGE!**

True or false?

a) 105·3 + 416·88 = 926 – 404·92

b) 2·6 + 13·13 + 8·4 = 102·87 – 77·74

## 3.14 Representing word problems involving decimal fractions

> We are learning to represent and solve the same word problem in different ways.

### Before we start

Finlay has saved £30·15 to buy these clothes for his holiday. He enters the prices into his calculator and gets the answer 35·8. Explain to Finlay what this means. How much does he still need to save?

 **£7·85**    **£12·46**   **£15·49**

> Representing the same problem in different ways helps to deepen our understanding.

### Let's learn

| Word problem | Bar model |
|---|---|
| The children are having a long jump competition. Nuria would need to jump a further 0·68 m to equal Isla's jump of 3·2 m. How far did Nuria jump? | Isla's jump: 3·2 m<br>Nuria's jump: ? \| 0·68 m |

**Answer**

**2·52 m**

**Empty Number Line**

−0·7

2·5   2·52      3·2
+0·02

**Jottings or algorithm**

$$
\begin{array}{r}
2\ ^{11}1 \\
\cancel{3}\cdot\cancel{2}0\,\text{m} \\
-\ 0\cdot68\,\text{m} \\
\hline
2\cdot52\,\text{m}
\end{array}
$$

**Let's practise**

Draw Think Boards like the one on the previous page to represent and solve these word problems.

1) In 1991, an American athlete, Mike Powell, jumped 8·95 m to set a world record. How much further is this than Isla's jump of 3·2 m?

2) The world record for the 200 m is 19·19 seconds. The world record for the 400 m is 43·03 seconds. Calculate the difference between these two times.

3) Nuria bought a computer game for £49·99, a magazine for £3·75 and a bag of sweets for 95p. She has £18·14 left. How much did Nuria have to start with?

4) Amman weighs 36·1 kg, his cat weighs 4·8 kg, his dog weighs 26·7 kg and his rabbit weighs 2·65 kg. What is their combined mass?

5) A forklift truck lifts three crates with a combined mass of 1016·7 kg. The first crate weighs 26·43 kg and the second crate weighs 57·8 kg. What is the mass of the third crate?

**CHALLENGE!**

Write your own word problems with at least two decimal fractions in them. Challenge a partner to draw Think Boards and solve your word problems.

## 3.15 Multi-step word problems

We are learning to solve multi-step word problems involving decimal fractions.

**Before we start**

In the 2018 European Championships, the British athlete ran 100 m in 9·95 seconds. The Turkish athlete ran 200 m in 19·76 seconds. Did the Turkish athlete take more or less than double the time to run double the distance?

When solving a challenging problem we need to think carefully about the strategies we will use and the steps we will take.

**Let's learn**

Finlay is working on this problem.

*Dad buys three burgers at £4·95 each, two portions of fries at £2·49 each, two milkshakes at £1·80 each and a cola.*

*He gives the cashier three £10 notes and receives £4·37 change. How much does the cola cost?*

He uses 'round and adjust' and place value to calculate the cost of the burgers, fries and milkshakes then adds the totals together.

$$(3 \times £5) - 15\,p = £14·85$$
$$(2 \times £2.50) - 2\,p = £\ 4·98$$
$$2 \times £1·80 \quad\quad = £\ 3·60$$
$$\overline{\phantom{(2 \times £2.50) - 2p =}\ £23·43}$$

(above £14·85 the small carried digits: 1 2 1)

Finlay draws a bar model to help him work out what to do next.

| £30 | | |
|---|---|---|
| £23·43 | £4·37 | ? |

$$£23·43$$
$$+£\ \ 4·37$$
$$\overline{£27·80}$$

£27·80 + **£2·20** = £30. The cola costs £2·20.

**Let's practise**

1) The world's tallest tropical tree measures 89·5 m. This is 740·3 m shorter than the world's tallest building and 86·99 m taller than the world's tallest man. What is the combined height of the tree, the building and the man?

2) The current world record for the hammer throw is 86·74 m. The world record for the javelin is 11·74 m further than the hammer. The world record for the shot put is 75·36 m less than that of the javelin. What is the current world record for the shot put?

3) Isla's mum spent £215·92 on a coat, a pair of shoes and a handbag. The coat cost £129·97. She paid £15·45 more for her shoes than for her handbag. How much did each item cost?

4) Four crates have a combined mass of 300 kg. Crate A weighs 54·7 kg. Crate B weighs half as much as Crate A and Crate C is three times heavier than Crate A. How heavy is Crate D?

**CHALLENGE!**

a) Find the combined mass of these creatures in kilograms.
b) How much heavier is the hippo than the hamster?

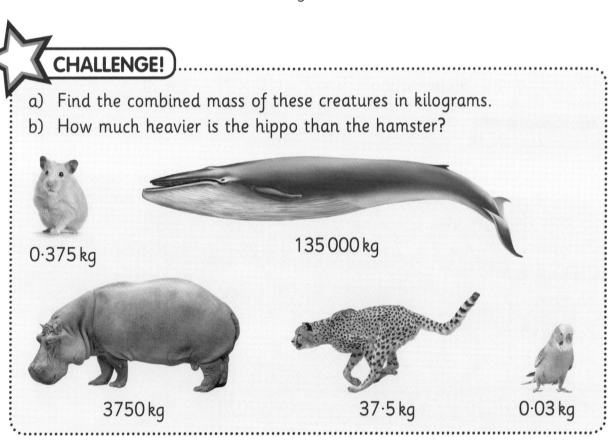

0·375 kg

135 000 kg

3750 kg

37·5 kg

0·03 kg

# 4 Number – multiplication and division

## 4.1 Multiplication and division facts for 7

We are learning to recall multiplication and division facts for 7.

**Before we start**

Isla is working out $112 \div 4$. How could she use her knowledge of multiplication facts for 4 to help her? Explain and work out the answer.

Recalling multiplication and division facts helps us use strategies and solve problems efficiently.

**Let's learn**

Multiples of 7 can be tricky to remember as they don't follow an obvious pattern. Try using knowledge of multiplication facts you have already to help, for example:

| | |
|---|---|
| Reversing the order | $3 \times 7 = 7 \times 3$ |
| Splitting the factors | $7 \times 7 = (5 \times 7) + (2 \times 7)$ |
| Rounding and compensating | $9 \times 7 = (10 \times 7) - (1 \times 7)$ |

**Let's practise**

1) How could you use multiplication facts you already know to work out these problems involving 7 if you didn't know them? An example has been done for you.

| | |
|---|---|
| a) $4 \times 7$ | $4 \times 7 = (2 \times 7) + (2 \times 7) = 28$ |
| b) $8 \times 7$ | |
| c) $6 \times 7$ | |
| d) $9 \times 7$ | |
| e) $7 \times 7$ | |
| f) $5 \times 7$ | |
| g) $11 \times 7$ | |

2) Work through these questions and think about which questions you recall quickly and which are more difficult to recall.

a) $7 \times \square = 21$  b) $7 \times 8 = \square$  c) $7 \times 3 = \square$

d) $35 \div 7 = \square$  e) $\square \times 7 = 63$  f) $14 \div 7 = \square$

g) $7 \times \square = 49$  h) $42 \div 7 = \square$  i) $7 \times \square = 70$

j) $56 \div 7 = \square$  k) $\square \div 7 = 4$  l) $63 \div 7 = \square$

Now list which facts you need to practise more.

3) Choose four of the 7 times table facts that you have to think about the most.

a) Complete four multiplication triangles based on them, as shown here.

Example:

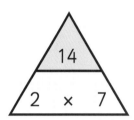

b) For each of your triangles, write a multiplication and a division fact.

c) Pass this book to a partner. Ask them to test you on the four 7 × table facts you found most difficult to recall.

## CHALLENGE!

Finlay thinks he has found another way to multiply numbers by 7. Firstly, he multiplies the number by 10. Then he multiplies the number by 3. Finally, he subtracts the second answer from the first.

For example:

$12 \times 10 = 120$

$12 \times 3 = 36$

$120 - 36 = 84$

So $12 \times 7$ must be 84.

Use this method to find out the answers to $11 \times 7$, $14 \times 7$, $15 \times 7$ and $20 \times 7$.

Does Finlay's method always work?

Can you come up with a quicker way to work out the answers?

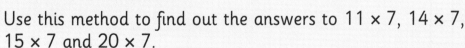

# 4 Number – multiplication and division

## 4.2 Recalling multiplication and division facts for 8

We are learning to recall multiplication and division facts for 8.

**Before we start**

Finlay knows the answer to 5 × 7. Show how he uses this to help him work out 5 × 14.

We can use facts we already know to help us learn multiplication and division facts for 8.

**Let's learn**

Multiplication facts for 2 and 4 can be very useful when multiplying numbers by 8.

If you know 6 × 2 = 12

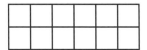

Then we can double the answer to work out 6 × 4 = 24

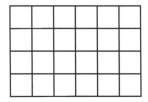

We can then double it again to work out 6 × 8 = 48

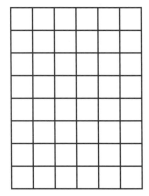

**Let's practise**

1) Nuria has used the double- double- double method to multiply 3 × 8 as shown:

|  | Double | Double | Double | So... |
|---|---|---|---|---|
| 3 × 8 | 3 × 2 = 6 | 6 × 2 = 12 | 12 × 2 = 24 | 3 × 8 = 24 |

Can you use the same method to work out these problems?

a) 5 × 8      b) 7 × 8          c) 2 × 8      d) 4 × 8

2) Write the multiples of 8 in 10 rows so that they make a pyramid.

The first three rows have been done for you.

Draw four coloured circles around the four multiples of 8 that you find hardest to recall quickly.

```
              8
          8       16
      8       16      24
```

3) Write one multiplication fact and one division fact for each of the numbers you circled in question 2.

**CHALLENGE!**

Nuria says that she can multiply numbers by 8 in the following way. She starts by multiplying the starting number by 10, which she finds easy to do.

Then she doubles the starting number. Finally, she subtracts the second answer from the first.

For example, to find the answer to 11 × 8:

11 × 10 = 110
11 × 2 = 22
110 − 22 = 88

So 11 × 8 must be 88.

Use this method to find out the answers to 12 × 8, 15 × 8, 18 × 8 and 20 × 8.

Can you explain why Nuria's method works?

See if you can find a quicker way to work out the answers.

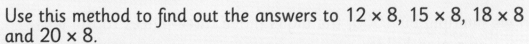

## 4.3 Multiplication with decimal fractions

We are learning to multiply decimal fractions by 10, 100 or 1000.

**Before we start**

There are 1000 g in each kg. How many grams are each of these measurements?

a) 23 kg      b) 5·2 kg      c) 16·1 kg
d) 0·9 kg      e) 468 kg      f) 1·1 kg

When we multiply whole numbers by 10, they get 10 times bigger.

**Let's learn**

Look at the place value table to see what happens when we multiply decimal fractions.

| HUNDREDS | TENS | ONES | TENTHS | HUNDREDTHS | |
|---|---|---|---|---|---|
| | | 0 | 2 | 4 | |
| | | 2 | 4 | | 0·24 × 10 |
| | 2 | 4 | 0 | | 0·24 × 100 |
| 2 | 4 | 0 | 0 | | 0·24 × 1000 |

When we multiply by 10, the digits move one place to the left and a zero is written in the empty column as a placeholder.

When we multiply by 100, the digits move two places to the left with zero as a placeholder.

When we multiply by 1000, the digits move three places to the left with zero as a placeholder.

Remember that the number gets bigger and the decimal point does not move. The decimal point is always fixed between the ones and the tenths.

**Let's practise**

1) Solve these problems:

    a) 2·28 × 10         b) 0·63 × 100         c) 18·46 × 100

    d) 14·55 × 1000      e) 5·02 × 100         f) 892·49 × 10

    g) 0·03 × 1000       h) 17·39 × 100

2) Answer true or false for each of these statements:

    a) 7·56 × 100 = 756         **True/False**

    b) 0·98 × 10 = 98           **True/False**

    c) 19·39 × 1000 = 193·9     **True/False**

    d) 189·27 × 1000 = 1892·7   **True/False**

    e) 0·02 × 10 = 0·2          **True/False**

    f) 254·99 × 100 = 254990    **True/False**

    g) 1000 × 8·26 = 8000·26    **True/False**

    h) 10 × 1453·42 = 14534·2   **True/False**

3) Nuria's pen has burst and made a mess of her homework. Can you help her by finding what number should be underneath the ink stains?

    a) 6·23 × ✳ = 62·3         b) 1·08 × ✳ = 108

    c) 99·99 × ✳ = 99990      d) ✳ × 1·64 = 164

    e) ✳ × 20·06 = 200·6      f) 0·999 × ✳ = 99·9

**CHALLENGE!**

The school office needs to order pencils for the school. It costs £3·19 for each pack of 10 pencils. They need 1000 altogether for the school.

How much will they need to spend?

## 4.4 Division with decimal fractions

We are learning to divide whole numbers by 10, 100 or 1000 with answers involving hundredths.

**Before we start**

How much is £6940 shared between 10 people?
How much is £6940 shared between 100 people?

When we divide whole numbers by 10, they get 10 times smaller.

**Let's learn**

Look at the place value table to see what happens when we divide decimal fractions.

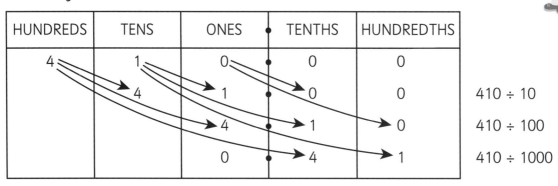

| HUNDREDS | TENS | ONES | • | TENTHS | HUNDREDTHS | |
|----------|------|------|---|--------|------------|---|
| 4 | 1 | 0 | • | 0 | 0 | |
| | 4 | 1 | • | 0 | 0 | 410 ÷ 10 |
| | | 4 | • | 1 | 0 | 410 ÷ 100 |
| | | 0 | • | 4 | 1 | 410 ÷ 1000 |

When a number is divided by 10, the digits move one place to the right. The same is true when the answer involves tenths and hundredths.

When a number is divided by 100, the digits move two places to the right, and when a number is divided by 1000, the digits move three places to the right. The same is true when the answer involves tenths and hundredths.

Remember that the number gets smaller and the decimal point does not move. The decimal point is always fixed between the ones and the tenths.

## Let's practise

1) Solve these questions:

    a) 52 ÷ 10            b) 6 ÷ 100           c) 412 ÷ 1000

    d) 82 ÷ 10            e) 99 ÷ 100          f) 12.5 ÷ 10

2) There are 10 millimetres in every centimetre, and 100 centimetres in every metre. Can you complete this table by converting the missing measurements?

| Millimetres | Centimetres | Metres |
|:---:|:---:|:---:|
| 1560 | 156 | |
| 67240 | | |
| | 989 | |
| | | 4930 |
| 114560 | | |

3) The building company needs to buy 1000 metres of wood. How much does it cost per metre at these different suppliers?

    a) £8920 for 1000 metres           b) £7990 for 1000 metres

    c) £10 010 for 1000 metres

### CHALLENGE!

Isla divides a number by 1000. Then she doubles it, and divides it by 10. Her answer is 0·28. Can you work out what number she started with?

Work with a partner to make up more problems like this for each other.

## 4.5 Solving multiplication problems

We are learning to multiply a four-digit number by a single-digit number and to solve these questions using partitioning by place value.

**Before we start**

It costs £100 each for the school trip. How much will it cost altogether for 653 children? Explain how you know.

We can partition numbers by place value to solve multiplication problems.

**Let's learn**

To make multiplying a four-digit number by a single-digit number easier, we can partition numbers into thousands, hundreds, tens and ones.

We can record this using the grid method, or by using brackets.

For example, to work out 2127 × 3:

| × | 2000 | 100 | 20 | 7 |
|---|------|-----|----|----|
| 3 | 6000 | 300 | 60 | 21 |
| 6000 + 300 + 60 + 21= 6381 | | | | |

$2127 \times 3 = (2000 \times 3) + (100 \times 3) + (20 \times 3) + (7 \times 3)$

$= 6000 + 300 + 60 + 21$

$= 6381$

**Let's practise**

1) Answer these questions using the grid method. The first one has been done for you.

a)  3186 × 9       b)  4562 × 8
c)  9365 × 4       d)  1873 × 6
e)  4294 × 7       f)  6651 × 5

| ×  | 3000   | 100 | 80  | 6  |
|----|--------|-----|-----|----|
| 9  | 27 000 | 900 | 720 | 54 |

27 000 + 900 + 720 + 54 = 28 674

2) Answer these questions using brackets. The first one has been done for you.

a)  7483 × 3 = (7000 × 3) + (400 × 3) + (80 × 3) + (3 × 3)
            = 21000 + 1200 + 240 + 9
            = 22 449
b)  4303 × 8       c)  5289 × 6        d)  2928 × 5
e)  4837 × 7       f)  1938 × 9

3) It costs £2863 per person for a holiday. How much would it cost for:

a) two people       b) seven people       c) nine people

Record your thinking using either the grid method or brackets.

**CHALLENGE!**

1) Isla ordered three boxes of 1075 exercise books for school. There are 3200 students in the school.
Did Isla order enough for one per student?   **yes / no**

Explain how you know this.

2) Isla ordered four boxes of 1685 pencils for school.
Did she order enough pencils so that each student can have two?
**yes / no**

Explain how you know this.

## 4.6 Solving multiplication problems involving decimal fractions

We are learning to multiply decimal fractions to two places by a single digit number.

**Before we start**

Look at this number:

How much does the 7 represent? How much does the 9 represent? What would the number one tenth more than this number be?

8·79

We can partition whole numbers by place value to make multiplying easier. We can use this strategy for decimal fractions, too.

**Let's learn**

Let's look at 4 × 5·46:

We could use the grid method to partition the number into hundredths, tenths and ones:

| × | 5 | 0·4 | 0·06 |
|---|----|-----|------|
| 4 | 20 | 1·6 | 0·24 |
| 20 + 1·6 + 0·24 = 21·84 | | | |

We could also use brackets to show how we have partitioned a problem:

4 × 5·46 = (4 × 5) + (4 × 0·4) + (4 × 0·06)

    = 20 + 1.6 + 0·24

    = 20 + 1·84

    = 21·84

**Let's practise**

1) Use the grid method to work out the answers to these problems. The first one has been done for you.

   a) 2 × 3·24       b) 6 × 1·27
   c) 4 × 1·8        d) 2·89 × 9
   e) 3 × 2·99

| × | 3 | 0·2 | 0·04 |
|---|---|-----|------|
| 2 | 6 | 0·4 | 0·08 |
| 6 + 0·4 + 0·08 = 6·48 | | | |

2) Use brackets to work out the answers to these problems. The first one has been done for you.

   a) 4 × 2·14 = (4 × 2) + (4 × 0·1) + (4 × 0·04)
              = 8 + 0·4 + 0·16
              = 8·56
   b) 5 × 9·38     c) 7 × 4·05      d) 7·16 × 2      e) 4 × 8·19

3) Isla is budgeting for a party. Can you work out the cost for each of these items?
   a) Three packets of balloons at £0·89 each.
   b) Five bottles of lemonade at £1·64 each.
   c) Four loaves of bread at £1·23 each.
   d) Eight packets of ham at £2·45 each.
   e) Six boxes of ice lollies at £3·97 each.

   Record your thinking using either the grid method or brackets.

**CHALLENGE!**

Work with a partner. Make up three problems for each other to answer that involve multiplying a decimal fraction to two places by a single-digit number. Work out your answers so you know if your partner gets it correct!

## 4.7 Solving division problems using place value

> We are learning to use partitioning by place value to solve division problems.

**Before we start**

Isla is reading a book. She reads six pages every day, and the book has 84 pages altogether. Can you use your knowledge of multiplication facts to help you work out how many days she takes to finish the book? Explain how you solved the problem.

> We can use partitioning numbers by place value to solve division problems.

**Let's learn**

Look at this grid method. We partition the three-digit number into hundreds, tens and ones:

For example, 225 ÷ 5:

We can also use multiplication facts we know already to partition numbers.

| ÷ | 200 | 20 | 5 |
|---|---|---|---|
| 5 | 40 | 4 | 1 |
| 40 + 4 + 1 = 45 | | | |

For example, 256 ÷ 4:

We know that 240 and 16 are both multiples of 4. We can then partition the 256 into 240 and 16 to make solving the problem easier:

256 ÷ 4 = (240 ÷ 4) + (16 ÷ 4)

          = 60 + 4

          = 64

**Let's practise**

1) Use the grid method to find your answers. The first one has been done for you.

a) 265 ÷ 5

| ÷ | 200 | 60 | 5 |
|---|-----|----|----|
| 5 | 40 | 12 | 1 |
| 40 + 12 + 1 = 53 | | | |

b) 695 ÷ 5

c) 524 ÷ 4

d) 378 ÷ 6

e) 432 ÷ 6

f) 976 ÷ 8

2) Partition these numbers so that you can use multiplication facts. Then answer the calculation.

a) 275 ÷ 5          b) 378 ÷ 3          c) 412 ÷ 4

d) 552 ÷ 6          e) 670 ÷ 5

3) Choose your own strategy to divide these numbers by 9 and complete the table.

a)

| 387 | 702 | 441 | 549 |
|-----|-----|-----|-----|
|     |     |     |     |

b) Choose your own strategy to divide these numbers by 8 and complete the table.

| 216 | 344 | 536 | 416 |
|-----|-----|-----|-----|
|     |     |     |     |

**CHALLENGE!**

Write the digits 0 to 9 on small pieces of paper. Find a place for each digit on these calculations so they are all correct:

a) ☐ 24 ÷ 4 = 2 ☐ 6

b) 723 ☐ ÷ 5 = ☐ 447

c) 3 ☐ 5 ☐ ÷ 6 = ☐ 59

d) 84 ☐ 5 ÷ ☐ = 1 ☐ 05

## 4.8 Solving multiplication problems using proportional adjustment

We are learning to use doubling and halving, or thirding and tripling to solve multiplication problems.

**Before we start**

Amman, Nuria and Isla all have a different strategy to solve the problem 7 × 18. Can you think of three different strategies they might have used, and use each of them to find the answer? Which strategy did you find the easiest?

We can double one factor and halve the other in a multiplication problem and the answer will not change.

**Let's learn**

Let's look at 16 × 5:

We can see that double each 5 makes 10, and that there are eight tens altogether:

16 × 5 = 8 × 10

**Half** of 16 is 8 and **double** 5 is 10.

When we double one factor and halve the other, the answer does not change.

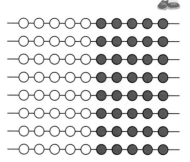

16 × 5

We can use this as a useful strategy to solve multiplication problems using facts we know.

This also works for thirding and tripling:

For example 3 × 27 = 9 × 9

**Triple** 3 is 9 and **a third of** 27 is 9.

We can triple one factor and third the other to solve the problem using facts we know.

1) Are these facts correct? Draw a diagram to justify your answer. The first one has been done for you. You can use cubes or counters to help you.

| a) 4 × 4 = 8 × 2 | ○ ○ ○ ○ ● ● ● ●<br>○ ○ ○ ○ ● ● ● ●<br>4 × 4 = 16<br>8 × 2 = 16 | Correct |
|---|---|---|
| b) 14 × 5 = 7 × 10 | | |
| c) 4 × 16 = 8 × 8 | | |
| d) 33 × 3 = 11 × 9 | | |
| e) 8 × 18 = 4 × 9 | | |

2) Find the answers to these using doubling and halving:

   a) 4 × 22    b) 3 × 16    c) 5 × 24
   d) 18 × 4    e) 28 × 5    f) 14 × 3

3) Could you use doubling and halving, or tripling and thirding, to help you solve these problems? Explain how you worked out each problem.

   a) Isla earned £12 a week for her paper round. How much had she earned altogether after seven weeks?
   b) Amman planted five rows of 42 strawberry plants. How many did he plant altogether?
   c) Finlay bought three shirts which each cost £24. How much did he spend altogether?

**CHALLENGE!**

Work with a partner and make up four problems for each other that might involve doubling and halving as a strategy. What do you notice when you are thinking of multiplication problems? Does this strategy work for all numbers? Explain why or why not.

## 4.9 Solving division problems

We are learning to divide using rounding and compensating.

**Before we start**

Amman is planting 48 bulbs in the garden. He is thinking how to organise the bulbs so that he has equal amounts of bulbs in each row. Draw arrays to show all the possible ways he could plant them to make sure each row is equal.

Rounding and compensating can be a useful strategy to make numbers in division problems easier to work with.

**Let's learn**

Let's look at $72 \div 4$:

We could round 72 up to 80 by adding eight ones:

We can then divide 80 by 4 giving us 20. Then we need to compensate – we divide the 8 we added by 4 and take this away from our total:

$72 \div 4 = (80 \div 4) - (8 \div 4)$

$= 20 - 2$

$= 18$

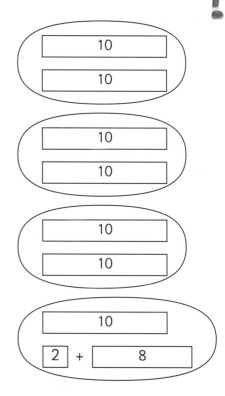

**Let's practise**

1) Solve the following division problems by rounding and compensating:

a) $54 \div 3 =$ [ $(60 \div 3) - (6 \div 3)$ ] $=$ [ ]

b) $95 \div 5 =$ [ ] $=$ [ ]

c) $96 \div 4 =$ [ ] $=$ [ ]

d) $76 \div 4 =$ [ ] $=$ [ ]

e) $144 \div 3 =$ [ ] $=$ [ ]

2) At the chair factory there are 96 wooden legs left. Each chair needs four legs. Use rounding and compensating to work out how many chairs can be made with 96 legs.

3) Isla takes £114 and spends the same amount each day on her holiday.

Can you use rounding and compensating to work out how many days it takes her to spend all her money if she spends these amounts every day:

a) £3 per day        b) £6 per day

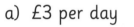

**CHALLENGE!**

Four friends won a competition with a prize of £192. Nuria uses rounding and compensating to work out how much they got each if they split the prize equally.

Could you show Nuria a different strategy to solve this problem?

Which strategy do you find the most efficient? Explain why.

# Number – multiplication and division

## 4.10 Solving multiplication problems

We are learning to multiply using written methods.

**Before we start**

Nuria has worked out the answer to the problem 28 × 5 as 108. Is she correct? Explain how you know.

We have worked on developing a range of mental strategies to solve multiplication problems in previous units, but it is also useful to know how to use written methods.

**Let's learn**

When multiplication problems are very challenging to solve mentally, written methods called algorithms can be very useful. An algorithm is a set of instructions that can be used to solve a problem.

Let's look at the problem 34 × 23.

In previous units we have learned how to use the grid method to partition by place value. We could use this here, like this:

| ×  | 30  | 4  |
|----|-----|----|
| 20 | 600 | 80 |
| 3  | 90  | 12 |
|    | 690 | 92 |
| 690 + 92 = 782 | | |

We could also use the standard algorithm for multiplication, where we set out the problem vertically by place value like this:

First, we multiply the top number by the ones.
Then we multiply the top number by the tens.
Then we add both numbers together vertically to get our total.

```
              34
            × 23
(34 × 3)     102
(34 × 20)    680
             782
```

**Let's practise**

1) Use the grid method to solve these problems. The first one has been set out for you:

a)  14 × 23

| × | 10 | 4 |
|---|----|---|
| 20 |   |   |
| 3 |   |   |
|   |   |   |

b)  31 × 45

c)  73 × 16

d)  52 × 28

2) Use the standard written method to solve these problems. The first one has been set out for you:

a)  28 × 33

|   | 2 | 8 |
|---|---|---|
| × | 3 | 3 |
|   |   |   |
|   |   |   |
|   |   |   |

(28 × 3)
(28 × 30)
Total

b)  19 × 23

c)  21 × 42

d)  27 × 19

3) Choose either the grid method or the standard written method to solve these problems:

a)  Fourteen containers each hold 82 ml. How much do they hold altogether?

b)  Each day Isla earns £15. How much has she earned after 28 days?

c)  The school hall can hold 27 rows of 35 chairs. How many chairs can the hall hold altogether?

**CHALLENGE!**

Here are three divisibility rules. One of them is false, but which one? Explain your answer.

- A number is divisible by 8 if the last 3 digits are divisible by 8.
- A number is divisible by 6 if the last 3 digits are divisible by 4.
- A number is divisible by 9 if the sum of its digits is divisible by 9.

## 4.11 Solving division problems using an algorithm

We are learning to divide using written methods.

### Before we start

Can you find three different ways to solve the problem 87 ÷ 3?
Share each strategy and explain which one you found most efficient for this problem and why.

We have worked on developing a range of mental strategies to solve division problems in previous units, but it is also useful to know how to use written methods.

### Let's learn

When division problems are difficult to solve mentally, we can use written methods to solve the problem.

Let's look at 92 ÷ 4.

When using written methods, we write this problem like this:

First, we divide the tens by 4. 90 ÷ 4 = **20** with a remainder of 10. We add the remaining 10 to the ones. Then we divide the remaining ones. There are 2 and the 10 remaining. 12 ÷ 4 = **3**

**92 ÷ 4 = 23**.

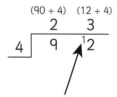

Add the 10 remaining to the ones.

### Let's practise

1) Use the written method to solve these division problems as shown:

a) 3 ) 57      b) 4 ) 68      c) 4 ) 104      d) 8 ) 96

2) Work out the answers to these problems using the written form:

a) 145 ÷ 5    b) 108 ÷ 6    c) 78 ÷ 3
d) 117 ÷ 9    e) 91 ÷ 7     f) 348 ÷ 3

3) There are 168 children on the school trip. Their teachers organise them into groups. Use the written method to work out how many groups will there be if there are:

a) four children in each group
b) six children in each group
c) seven children in each group

CHALLENGE!

1) Play the game, Division Squares, with a partner using the board on this page. One plays with red counters, the other with blue.

- Choose a different colour counter to be your playing piece and start by placing it on any white square.

- Spin a 0–3 spinner and move your counter that number of spaces in any horizontal or vertical direction.

- If you land on a division problem, answer the question.

- If your answer is correct, place a counter of your colour on that square and spin again.

- When all division problems are covered, the player with the most counters on the board is the winner.

Think about all the strategies you have learned so far, mental and written, and use the one you feel is most efficient for solving each question.

| | | | 909 ÷ 3 | | | |
|---|---|---|---|---|---|---|
| 372 ÷ 6 | | 854 ÷ 7 | | 976 ÷ 8 | | |
| | | | | | 429 ÷ 3 | |
| 915 ÷ 5 | | 584 ÷ 8 | | 870 ÷ 6 | | |
| | | | | | | |
| 441 ÷ 7 | | 728 ÷ 4 | | 585 ÷ 5 | | 738 ÷ 9 |
| | | | | | | |
| | | 954 ÷ 9 | | | | 656 ÷ 4 |

## 4.12 Solving division problems involving decimal fractions

We are learning to divide decimal fractions to two places by a single-digit number.

**Before we start**

Draw a diagram to show the value of each of these numbers:
a) 3·17     b) 0·02     c) 1·99     d) 2·56

We can use written methods to divide decimal fractions.

**Let's learn**

Let's look at 4·32 ÷ 3:

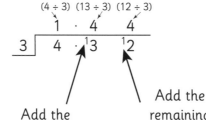

(4 ÷ 3)  (13 ÷ 3)  (12 ÷ 3)

Add the remaining 1 to the tenths giving 13 tenths

Add the remaining 1 tenth to the hundredths giving 12 hundredths

First, we divide the ones by 3.

**4 ÷ 3 = 1 with a remainder of 1**

We add the remaining 1 to the tenths. Then we divide the tenths. There are 1 whole remaining and 3 tenths which gives us 13 tenths.

**13 ÷ 3 = 4 with a remainder of 1**

We add this one tenth to the hundredths, which gives us 12 hundredths.

**12 ÷ 3 = 4**

**4·32 ÷ 3 = 1·44**

**Let's practise**

1) Use the written method to work out the answers to these problems.

   a) $4\overline{)1\cdot24}$     b) $3\overline{)6\cdot57}$     c) $8\overline{)1\cdot04}$     d) $6\overline{)18\cdot54}$

2) Work out the answers to these problems using the written form:

   a) $4\cdot75 \div 5$         b) $5\cdot12 \div 8$         c) $2\cdot88 \div 3$
   d) $4\cdot41 \div 7$         e) $2\cdot64 \div 4$         f) $6\cdot39 \div 9$

3) Nuria went shopping. Here is a copy of her receipt:

| North West Corner Shop | |
|---|---|
| 6 apples | £ 2·82 |
| 4 chocolate bars | £ 3·92 |
| 7 oranges | £ 2·24 |
| 5 packets of biscuits | £ 6·30 |
| 9 muesli bars | £ 7·83 |
| 3 bottles of juice | £ 4·68 |

Can you work out how much each individual item cost?

   a) one apple          b) one chocolate bar
   c) one orange         d) one packet of biscuits
   e) one muesli bar     f) one bottle of juice

**CHALLENGE!**

Finlay and Isla are solving the problem $4\cdot16 \div 4$. Finlay's answer is 4·4 and Isla's answer is 4·04. Who is correct?

Can you think of three different strategies to solve this problem?

Which way did you think was the most efficient? Explain why.

## 4.13 Solving multi-step problems using the order of operations

> We are learning to solve problems that involve all four operations.

**Before we start**

> Isla is going to the cinema with three friends. They have £38 to spend between them and it costs £8 each for a ticket. How much money will they have left to spend on snacks?

> When we are solving problems that involve a combination of addition, subtraction, multiplication and division, the order that we use for each calculation is important.

**Let's learn**

When problems involve all four operations, we work out the multiplication and division calculations first and then the addition and subtraction calculations. We call this the **order of operations.**

For example, if we were working out this problem:

$6 + 5 \times 8 - 3$

We start from the left and work to the right. Then we look for multiplication or division calculations first. Multiplication and division have equal importance.

Then we work out addition or subtraction calculations. Addition and subtraction have equal importance.

$6 + (5 \times 8) - 3$

$= 6 + 40 - 3$

$= 43$

> We learned in previous units that we use brackets to show which calculations we do first.

## Let's practise

1) Find the answer to these problems using the order of operations. Use brackets to show how you solved each problem.

   a) $4 \times 3 - 6$

   b) $21 - 2 \times 9 + 5$

   c) $9 \times 2 + 8 \div 4$

   d) $42 \div 7 - 3 + 16$

   e) $4 + 27 \div 3 - 4$

   f) $25 \div 5 + 18 \div 6$

   g) $7 \times 4 + 9 \times 2 - 5$

   h) $16 - 6 \div 2 + 5 \times 4$

2) Finlay is worked out these problems as shown. Will he get the correct answer? Write yes or no. If he is incorrect, can you help him by showing how to use the order of operations to solve the problem?

   a) $6 + 14 \div 2 = (6 + 14) \div 2$

   b) $5 \times 6 - 2 \times 3 = (5 \times 6) - (2 \times 3)$

   c) $14 - 4 \div 2 + 8 \times 4 = (14 - 4) \div 2 + (8 \times 4)$

   d) $48 \div 8 - 4 + 5 \times 2 = (48 \div 8) - (4 + 5) \times 2$

3) Nuria goes shopping and buys three books which each cost £15 and another two which cost £8 each. Work out how much change she has left if she started with these amounts:

   a) £70

   b) £100

   c) £85

### CHALLENGE!

Use all the digits given to make a number sentence that involves multiplication (don't forget the equals symbol and the answer!). Then write a word problem for one sentence in section 1 and one in section 2 that can be solved using that number sentence.

1) Two-digit by one-digit multiplication:

   a) 3 9 3 9 3

   b) 9 9 9 9 1 8

   c) 0 2 4 1 5 5

2) Three-digit by one-digit multiplication:

   a) 4 1 4 1 4 1 4

   b) 5 5 7 5 2 7 5 5

   c) 3 3 0 0 1 3 9

# 5 Multiples, factors and primes

## 5.1 Using knowledge of multiples and factors to work out divisibility rules

**Before we start**

Nuria says that the number 12 has four factors. Is she correct? Explain your answer.

We are learning to see if a number can be divided exactly by another number.

It is useful to work out simple rules to help us quickly and easily see if a number can be divided exactly by another number. We call these **divisibility rules.**

**Let's learn**

Think about what you know about multiples of 2. What do they all have in common?

We can see that they are all **even numbers.** Therefore, we know that to be divisible by 2, a number would have to be an even number. This is a divisibility rule for 2.

Let's look at divisibility rules for 3. Is 621 divisible by 3?

To test if a number can be exactly divided by 3 we add all the digits of a number together. This is called the **reduced number**. If the sum of the digits is a multiple of 3, we know that the number is divisible by 3.

6 + 2 + 1 = 9 and 9 is a multiple of 3 so 621 **is** divisible by 3.

This rule works for 9, too. Find the reduced number to see if it is a multiple of 9. Is 739 divisible by 9?

7 + 3 + 9 = 19. 19 is not a multiple of 9, so 739 **is not** divisible by 9.

What about divisibility rules for 4? Is 816 divisible by 4?

To test if a number can be divided exactly by 4, we look at the last two digits of the number.

16 ÷ 4 = 4, so 816 **is** divisible by 4.

Use your knowledge of these rules and multiples of other numbers to work out these problems.

**Let's practise**

1) a) Are these numbers divisible by 5? Circle all the numbers that can be divided exactly by 5. You can use a calculator to help.

| 75 | 102 | 890 | 524 | 47 855 | 7540 | 7386 |

| 890 635 | 182 780 | 38 984 | 748 905 |

b) Can you work out a divisibility rule for 5? Write it here.

2) Think about multiples of 10. Write a divisibility rule for 10 and explain why it works.

3) Use your knowledge of multiples and divisibility rules to work out if these answers are true or false:

a) 319 is divisible by 3.    **True/False**
b) 524 is divisible by 4.    **True/False**
c) 828 is divisible by 9.    **True/False**
d) 5974 is divisible by 5.    **True/False**
e) 1714 is divisible by 4.    **True/False**
f) 2772 is divisible by 9.    **True/False**
g) 475 is divisible by 3.    **True/False**
h) 356 890 is divisible by 10.    **True/False**

**CHALLENGE!**

Look at these numbers:

| 724 | 126 | 625 | 987 | 1026 | 312 | 782 |

| 216 | 684 | 1524 | 372 | 528 |

a) Circle all the numbers that are divisible by 6. You can use a calculator to help if you need to. What do you notice?

b) Can you work out a divisibility rule for 6?
   Hint: Try finding the reduced number.

c) How about a divisibility rule for 8?

## 5.2 Using knowledge of multiples and factors to solve problems

We are learning to use what we know about multiples and factors to solve problems.

**Before we start**

Write three different numbers that are multiples of 2, 5 **and** 10. What do you notice about your answers?

Our knowledge of factors and multiples can be very useful when we are solving problems!

**Let's learn**

When we are working out problems, we can use what we know about multiples and factors to help us solve them efficiently.

For example, if we had 24 bulbs to plant in the garden and we wanted to plant them in equal rows, we could use our knowledge of the factors of 24.

If we know that 1, 2, 3, 4, 6, 8, 12 and 24 are all factors of 24, then we can quickly work out all the different ways we could plant each row to help us make our decision.

Our options would be two rows of 12, three rows of 8, four rows of 6 or one row of 24.

Knowledge of multiples and factors can also help us check if our answers to multiplication or division problems are reasonable.

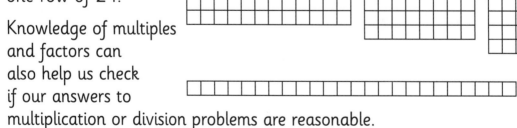

For example, if we work out 17 × 4 as 55, we can quickly see that our answer must be incorrect if we know that all multiples of 4 are even numbers.

**Let's practise**

1) Finlay is helping out at the farm. There are 315 eggs that all need to be sorted into half-dozen boxes (boxes of six).

   a) Will the eggs all fit into the boxes with none left over? How do you know?

   b) How many boxes will Finlay need altogether?

   c) If he only had boxes that held a dozen eggs, how many would he need?

2) Everyone at Isla's school has made a decorative tile for the wall. There are 64 tiles altogether. The headteacher wants them to be put on the wall in reception to make a rectangular shape. Can you use your knowledge of factors to work out all the possible ways they could be placed in arrays?

3) Nuria and Amman have worked out these calculations with the answers shown. Use your knowledge of factors and multiples to estimate who is correct. Explain how you know.

   The first one has been done for you.

| Problem | Nuria's answer | Amman's answer | Who is correct? |
|---------|---------------|----------------|-----------------|
| 38 × 10 | 380 | 76 | Nuria. Multiples of 10 always end in a zero, so Amman's answer can't be correct. |
| 73 × 5 | 438 | 365 | |
| 155 ÷ 2 | 75 | 77·5 | |
| 26 × 8 | 208 | 64 | |
| 36 × 9 | 324 | 360 | |

**CHALLENGE!**

Isla says her dad's age is a multiple of 7, but after his next birthday it will be a multiple of 6. Can you work out how old he is?

Can you make up more problems like this for a partner to solve?

# 6 Fractions, decimal fractions and percentages

## 6.1 Converting fractions

We are learning to convert mixed numbers to improper fractions.

**Before we start**

Simplify the following: $\frac{14}{20}$ $\frac{8}{32}$ $\frac{35}{40}$ $\frac{7}{10}$

Fractions greater than 1 can be expressed as proper and improper fractions.

**Let's learn**

Finlay has two whole pizzas and one half pizza.

How much pizza do I have here?

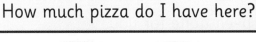

We can also say that he has five half pizzas.

We can write this as:

- a **mixed number**: $2\frac{1}{2}$ pizzas (using a mix of whole numbers and fractions).

  Or

- an **improper fraction**: $\frac{5}{2}$ (using only fractions).

We can use a bar model to convert between mixed numbers and improper fractions:

Mixed number: | one whole | one whole | one half | $2\frac{1}{2}$

Each whole can be split into two halves:

Improper fraction: | one half | one half | one half | one half | one half | $\frac{5}{2}$

## Let's practise

1) Convert each of these improper fractions into a mixed number (the first one has been done for you):

a) $\frac{5}{3}$ = $2\frac{2}{3}$

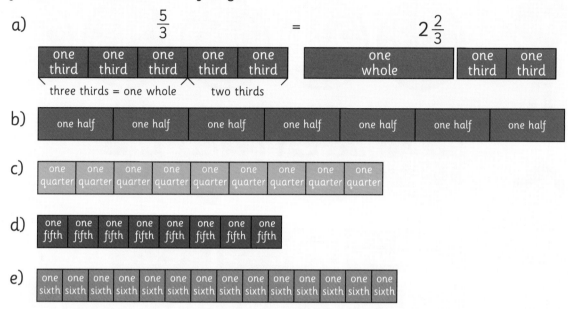

b)

c)

d)

e)

2) Convert each of these mixed numbers into an improper fraction (the first one has been done for you):

a) $1\frac{3}{4}$ = $\frac{7}{4}$

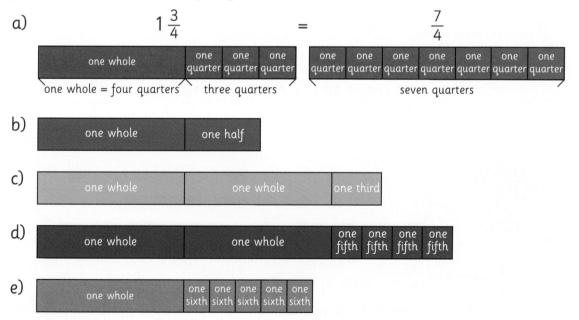

b)

c)

d)

e)

3) Write each of the children's portions as a mixed number and improper fraction:

a)

b)

c)

d)

CHALLENGE!

1) Amman has been asked to put these fraction cards in order from largest to smallest. Can you help him?

| $\frac{16}{5}$ | $\frac{10}{3}$ | $\frac{35}{10}$ | $\frac{19}{6}$ | $\frac{51}{15}$ |

I don't know how to compare these because they all have different denominators.

2) Can you find a fraction that fits in between each of these pairs of fractions?

| $\frac{7}{3}$ | $\frac{5}{2}$ | | $\frac{11}{3}$ | $\frac{15}{4}$ | | $\frac{8}{5}$ | $\frac{7}{4}$ |

# 6 Fractions, decimal fractions and percentages

## 6.2 Comparing and ordering fractions

We are learning to compare and order any fraction (including improper fractions).

### Before we start

Calculate an equivalent fraction for each of the following and order them from smallest to largest:

$\frac{2}{3}$ $\qquad$ $\frac{7}{12}$ $\qquad$ $\frac{3}{4}$ $\qquad$ $\frac{5}{8}$

We can compare any fractions by finding a common equivalent.

### Let's learn

I can't convert halves into fifths or fifths into halves, so I'm not sure how to compare these.

Would you rather get half or three-fifths of a bag of sweets?

In order to compare these fractions, the children will need to convert them both into a **common equivalent** fraction. Halves and fifths can both be converted into tenths:

We'll need to find a fraction that both can be converted into.

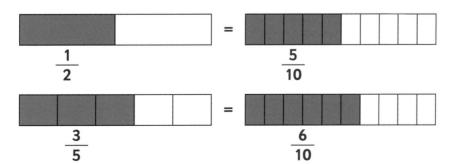

$$\frac{1}{2} \quad = \quad \frac{5}{10}$$

$$\frac{3}{5} \quad = \quad \frac{6}{10}$$

Once they have been converted to tenths, we can easily compare. Six tenths is greater than five tenths so three fifths is greater than one half.

Could we have found any other common equivalents so that we could compare one half with three-fifths?

## Let's practise

1) Write a common equivalent to help you solve the following:

a) Nuria has read two-thirds of a book. Isla has read three-quarters of the same book. Who has read more?

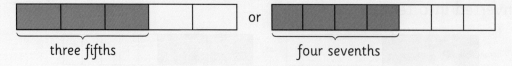

two thirds     or     three quarters

b) Isla and Amman have identical packs of modelling clay. Isla uses three-eighths of her clay and Amman uses five-twelfths of his. Who uses more of their clay?

three eighths     or     five twelfths

c) Amman has three-fifths of a bottle of juice. Finlay has four-sevenths of an identical bottle. Who has more juice?

three fifths     or     four sevenths

2) Use equivalence to compare the following sets of fractions (the first one has been done for you):

a)  $\boxed{\dfrac{5}{2}}$  and  $\boxed{\dfrac{8}{3}}$

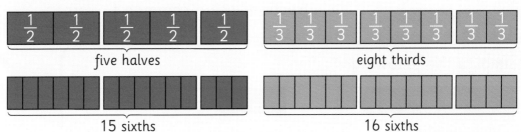

five halves          eight thirds

15 sixths          16 sixths

$\dfrac{15}{6}$ is less than $\dfrac{16}{6}$  so  $\dfrac{5}{2}$ is less than $\dfrac{8}{3}$

b)  $\boxed{\dfrac{7}{2}}$  and  $\boxed{\dfrac{15}{4}}$     c)  $\boxed{\dfrac{5}{3}}$  and  $\boxed{\dfrac{7}{4}}$

d)  $\boxed{\dfrac{12}{5}}$  and  $\boxed{\dfrac{9}{4}}$     e)  $\boxed{\dfrac{11}{3}}$  and  $\boxed{\dfrac{17}{5}}$

## CHALLENGE!

The children are taking part in a basketball competition and want to compare their scores:

I scored 24 baskets out of 32 attempts.

I scored 28 baskets out of 56 attempts.

I scored 22 baskets out of 33 attempts.

I scored 15 baskets out of 25 attempts.

Can you help them find a way to compare their scores fairly?

Write the children's names in order from first to last.

# 6 Fractions, decimal fractions and percentages

## 6.3 Simplifying fractions

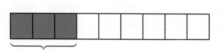

We are learning to simplify fractions using common factors.

**Before we start**

Simplify the following:

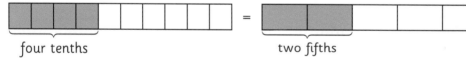

eight tenths      three ninths

We can use common factors to find a fraction in its simplest form.

**Let's learn**

A fraction can be simplified by splitting it into fewer parts.

four tenths   =   two fifths

It takes less effort to cut the bar into fifths than cutting it into tenths. It's simpler to make!

$\frac{4}{10}$ can be **simplified** to make $\frac{2}{5}$

We can **simplify** fractions using **common factors**:

2 is a factor of both the 4 and 10. We say that 2 is a **common factor** of both.

Instead of needing to cut the whole bar into 10 parts, we only need to cut it into five parts.

Instead of getting four of the tenths, we are getting two of the fifths.

$\frac{2}{5}$ is **equivalent** to $\frac{4}{10}$ .

number of parts we are getting

$$\frac{4}{10} \xrightarrow[\div 2]{\div 2} = \frac{2}{5}$$

number of parts that make up the 'whole'

**Let's practise**

1) Use common factors to calculate an equivalent fraction in its simplest form for each of the following. Draw each fraction in its simplest form

a) 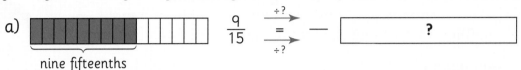 $\dfrac{9}{15}$ $\begin{array}{c} \xrightarrow{\div?} \\ = \\ \xrightarrow{\div?} \end{array}$ — [ ? ]

nine fifteenths

b) $\dfrac{55}{80}$ $\begin{array}{c} \xrightarrow{\div?} \\ = \\ \xrightarrow{\div?} \end{array}$ — [ ? ]

55 eightieths

c) $\dfrac{14}{35}$ $\begin{array}{c} \xrightarrow{\div?} \\ = \\ \xrightarrow{\div?} \end{array}$ — [ ? ]

14 thirty-fifths

d) $\dfrac{30}{36}$ $\begin{array}{c} \xrightarrow{\div?} \\ = \\ \xrightarrow{\div?} \end{array}$ — [ ? ]

30 thirty-sixths

2) Isla finds a fraction in its simplest form in steps - i) find a common factor; ii) simplify the fraction; iii) check to see if it can be simplified further.

Use Isla's strategy to simplify the following (the first one has been done for you):

a) $\dfrac{50}{100} \begin{array}{c} \xrightarrow{\div 2} \\ = \\ \xrightarrow{\div 2} \end{array} \dfrac{25}{50} \begin{array}{c} \xrightarrow{\div 5} \\ = \\ \xrightarrow{\div 5} \end{array} \dfrac{5}{10} \begin{array}{c} \xrightarrow{\div 5} \\ = \\ \xrightarrow{\div 5} \end{array} \dfrac{1}{2}$

b) $\dfrac{60}{100} \begin{array}{c} \xrightarrow{\div?} \\ = \\ \xrightarrow{\div?} \end{array} — \begin{array}{c} \xrightarrow{\div?} \\ = \\ \xrightarrow{\div?} \end{array} — \begin{array}{c} \xrightarrow{\div?} \\ = \\ \xrightarrow{\div?} \end{array} —$

c) $\dfrac{32}{48} \begin{array}{c} \xrightarrow{\div?} \\ = \\ \xrightarrow{\div?} \end{array} — \begin{array}{c} \xrightarrow{\div?} \\ = \\ \xrightarrow{\div?} \end{array} — \begin{array}{c} \xrightarrow{\div?} \\ = \\ \xrightarrow{\div?} \end{array} —$

d) $\dfrac{90}{150} \begin{array}{c} \xrightarrow{\div?} \\ = \\ \xrightarrow{\div?} \end{array} — \begin{array}{c} \xrightarrow{\div?} \\ = \\ \xrightarrow{\div?} \end{array} — \begin{array}{c} \xrightarrow{\div?} \\ = \\ \xrightarrow{\div?} \end{array} —$

**CHALLENGE!**

Choose one of the fractions below and see how many different ways you can simplify it using common factors:

i) $\dfrac{45}{60}$    ii) $\dfrac{48}{72}$    iii) $\dfrac{70}{100}$    iv) $\dfrac{30}{50}$

a) Can you calculate each fraction in its simplest form?

b) Which of the fractions above has the most common factors?

# 6 Fractions, decimal fractions and percentages

## 6.4 Adding and subtracting fractions

We are learning to add and subtract fractions using equivalence.

### Before we start

Identify three fractions that are equivalent to each of the following:

$\frac{2}{3}$    $\frac{3}{4}$

We can use equivalence to add and subtract fractions that are not equivalent.

### Let's learn

 Amman has half an orange and Finlay has a quarter portion. They're trying to work out how much they have altogether.

I think we have two halves altogether.

I think we have two quarters.

 $\frac{1}{2} + \frac{1}{4} = \frac{2}{6}$

I think you have two sixths.

Amman can't be right because they don't have as much as two halves altogether.

Finlay can't be right because they have more than two quarters altogether.

Let's use a bar model to see if Nuria is correct:

| one half | | + | one quarter | | | | $\neq$ | one sixth | one sixth | | | |

One half plus one quarter is clearly much bigger than two sixths so Nuria can't be right either.

| one half | | = | one quarter | one quarter | | |

In order to add these portions we need to convert them into like fractions:

Now we can add the portions together:

 +  +

2 quarters        1 quarter        3 quarters

**Let's practise**

1) Use equivalence to solve the following:

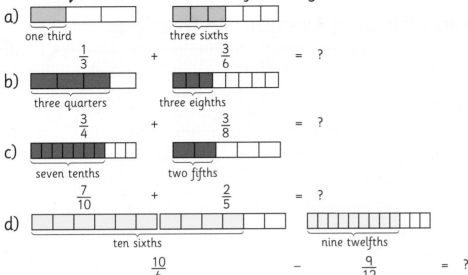

a) one third $\frac{1}{3}$ + three sixths $\frac{3}{6}$ = ?

b) three quarters $\frac{3}{4}$ + three eighths $\frac{3}{8}$ = ?

c) seven tenths $\frac{7}{10}$ + two fifths $\frac{2}{5}$ = ?

d) ten sixths $\frac{10}{6}$ – nine twelfths $\frac{9}{12}$ = ?

2) Draw bar models to help solve the following:

a) $\frac{3}{10} + \frac{4}{5} =$ ____  b) $\frac{11}{8} - \frac{3}{4} =$ ____  c) $\frac{2}{3} + \frac{9}{12} =$ ____  d) $2\frac{1}{2} - \frac{5}{6} =$ ____

3) Draw diagrams to solve the following problems:

a) Finlay runs $1\frac{1}{4}$ miles and then cycles $1\frac{11}{12}$ miles. What is the total distance he travels?

b) Amman is carrying a bag of shopping that weighs $2\frac{3}{10}$ kilograms. He removes a bag of potatoes weighing $\frac{4}{5}$ of a kilogram. What does the shopping bag weigh now?

c) Isla is walking the $4\frac{1}{6}$-mile path to the top of Ben Nevis. So far, she has walked $1\frac{2}{3}$ miles. How far does she still have to go to reach the top?

**CHALLENGE!**

a) Isla is trying to work out what fraction of children in her school have brown eyes. She knows that half of the children have blue eyes, one third of the children have green eyes, and the rest of the children have brown eyes.

There are 330 children in the school. How many children have brown eyes?

b) Amman eats one fifth of a bag of sweets. Finlay eats one quarter of the bag. What fraction of sweets is left in the bag?

# 6 Fractions, decimal fractions and percentages

## 6.5 Converting decimal fractions to fractions

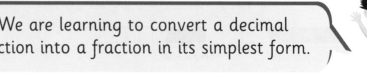

We are learning to convert a decimal fraction into a fraction in its simplest form.

**Before we start**

Convert the following decimal fractions into fractions:

0·3          0·55          0·79          0·8

Decimal fractions can be converted into fractions. Same value, different appearance!

**Let's learn**

We know that the first place after the decimal point represents tenths and the second place represents hundredths:

**0·23 =**     **2 tenths**          and          **3 hundredths**

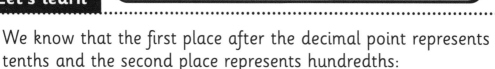

Or:

**0·23 =**     **23 hundredths**

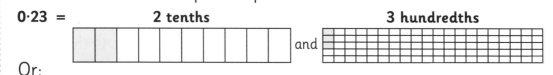

We can convert any decimal fraction into a fraction:

**0·25 =**     **25 hundredths**          $= \dfrac{25}{100}$

We can simplify this fraction using equivalence. In order to create 25 hundredths, we need to split the bar into 100 equal parts and colour 25 of the parts. We can find exactly the same portion of the bar if we split it into four equal parts and colour one of the parts:

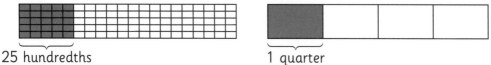

25 hundredths                    1 quarter

So,

**0·25 =**          $\dfrac{25}{100}$          =          $\dfrac{1}{4}$

## Let's practise

1) Match the following fractions to their decimal equivalents:

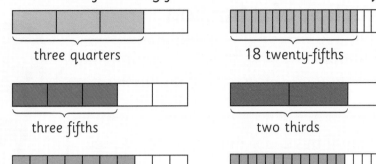

three quarters      18 twenty-fifths     0·65   0·7

three fifths        two thirds       0·72   0·67

seven tenths      13 twentieths     0·75   0·6

2) Write the following decimal fractions as a fraction in its simplest form

a) **0·75** = [＿＿＿＿] = [＿＿＿＿＿＿＿] = ?
            ?                ?

b) **0·35** = [＿＿＿＿] = [＿＿＿＿＿＿＿] = ?
            ?                ?

c) **0·12** = [＿＿＿＿] = [＿＿＿＿＿＿＿] = ?
            ?                ?

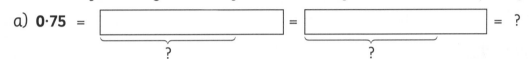

### CHALLENGE!

Nuria has challenged Isla to convert each of these decimal numbers to a fraction in its simplest form:

a) 0·88    b) 1·45    c) 3·9     d) 5·72    e) 4·6

Help Isla to complete Nuria's challenge. Show your answers as fractions and bar diagrams.

# 6 Fractions, decimal fractions and percentages

## 6.6 Calculating a fraction of a fraction

We are learning to solve problems by finding a fraction of a fraction.

**Before we start**

Demonstrate how we can solve this problem using a bar model:

Two-thirds of the fans at a rugby match support the red team. There are 360 fans at the match in total. How many fans of the red team are at the match?

Bar models help us visualise problems where we need to find a fraction of a fraction.

**Let's learn**

Isla has half a bag of sweets. She gives one third of her sweets to Nuria. What fraction of a bag of sweets does Nuria have?

We can use a bar model to solve this problem:

whole bag { one half / one half } Isla starts with

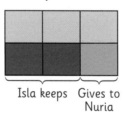

Isla keeps    Gives to Nuria

When we split one half into three equal parts, we can see that one of those parts is equivalent to a sixth of the whole bag.

We can say that:

One third of a half-bag of sweets is one sixth of a bag of sweets.

Or

$$\frac{1}{3} \times \frac{1}{2} = \frac{1}{6}$$

1) Draw a bar model to solve each of the following:

a) What is half of one sixth of a bar?

b) What is half of three-quarters of a bar?

c) What is two-thirds of half a bar?

d) What is one quarter of two-thirds of a bar?

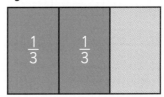

2) Draw a bar model to solve each of the following problems:

a) Finlay has half a bag of sweets. He gives one quarter of his sweets to Amman. What fraction of a bag of sweets does Amman have?

b) Isla cuts herself one third of a cake. She eats one third of her slice. What fraction of the whole cake has she eaten?

c) Nuria has one third of a bottle of water. She spills half of her water. What fraction of the whole bottle does she have left?

**CHALLENGE!**

Amman spends three-fifths of his birthday money and puts the rest of it in the bank.

- Half of the money **he spent** was on a pair of trainers.
- One third of the money **he spent** was on a football top.
- One sixth of the money **he spent** was on a set of headphones.

What fraction of his **birthday money** was spent on:

a) the pair of trainers  b) the football top  c) the set of headphones

# 6 Fractions, decimal fractions and percentages

## 6.7 Dividing a fraction by a whole number

> We are learning to divide a proper fraction by a whole number.

### Before we start

Write a word problem that could be represented by this bar model and solve it:

32

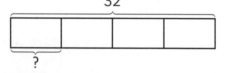

?

> Fractions can be divided by splitting them into equal parts.

### Let's learn

Fractions can be larger or smaller than one.

Isla has half a bar of chocolate and Finlay has one and a half bars of chocolate.

We can write these fractions as $\frac{1}{2}$ and $\frac{3}{2}$.

A portion that is less than one whole is called a **proper fraction**.

A portion that is greater than one whole is called an **improper fraction**.

When we divide a fraction into equal parts, we get more parts but each part is smaller.

Let's try dividing Isla's half portion into two parts to share with Amman:

 shared between two people

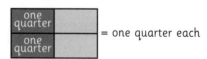 = one quarter each

When we divide Isla's portion into two parts, we create quarters. Isla and Amman would each get one quarter of a chocolate bar.

We can say that: $\frac{1}{2} \div 2 = \frac{1}{4}$

## Let's practise

1) Draw a bar model to solve how the following could be shared out equally:

a) $\frac{1}{4}$ shared between three people

| ? |
|---|

÷ = each

b) $\frac{1}{5}$ shared between four people

| ? |
|---|

÷ = each

c) $\frac{1}{3}$ shared between four people

| ? |
|---|

÷ = each

2) Draw a bar model to solve how the following could be shared out equally:

a) $\frac{1}{4}$ $\frac{1}{4}$ $\frac{1}{4}$ shared between four people

| ? |
|---|

÷ = each

b) $\frac{1}{5}$ $\frac{1}{5}$ $\frac{1}{5}$ shared between two people

| ? |
|---|

÷ = each

c) $\frac{1}{6}$ $\frac{1}{6}$ $\frac{1}{6}$ $\frac{1}{6}$ $\frac{1}{6}$ shared between three people

| ? |
|---|

÷ = each

## CHALLENGE!

The children are baking cakes for the school fayre and each have some cups of flour:

I have $6\frac{1}{2}$ cups of flour.

I have $3\frac{1}{2}$ cups of flour.

I have $5\frac{1}{3}$ cups of flour.

I have $4\frac{1}{4}$ cups of flour.

- Amman needs half a cup of flour to make one muffin.
- Nuria needs one quarter of a cup of flour to make one pancake.
- Finlay needs one sixth of a cup of flour to make one scone.
- Isla needs one eighth of a cup of flour to make one cupcake.

How many cakes can each of the children make?

## 6.8 Dividing a whole number by a fraction

> We are learning to divide a whole number by a fraction.

**Before we start**

Write a word problem for the following bar model and then solve it:

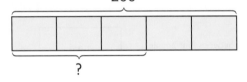

200

?

> When we divide an amount by a proper fraction the answer is greater than the original amount.

**Let's learn**

Nuria and Amman have been asked to share out boxes of pencils to the class:

> How many children can we give half a box to?

> There are 10 boxes and each child is to get half a box.

| pencils | pencils | pencils | pencils | pencils | pencils | pencils | pencils | pencils | pencils |

We can share the 10 boxes into half portions:

| $\frac{1}{2}$ $\frac{1}{2}$ | $\frac{1}{2}$ $\frac{1}{2}$ | $\frac{1}{2}$ $\frac{1}{2}$ | $\frac{1}{2}$ $\frac{1}{2}$ | $\frac{1}{2}$ $\frac{1}{2}$ | $\frac{1}{2}$ $\frac{1}{2}$ | $\frac{1}{2}$ $\frac{1}{2}$ | $\frac{1}{2}$ $\frac{1}{2}$ | $\frac{1}{2}$ $\frac{1}{2}$ | $\frac{1}{2}$ $\frac{1}{2}$ |

Ten boxes divided into half a box each means we have enough to give pencils to 20 children. We can write:

$$\text{10 boxes} \div \frac{1}{2} \text{ portions = 20 pupils} \quad \textbf{OR} \quad 10 \div \frac{1}{2} = 20$$

## Let's practise

1) The children are preparing for a buffet by cutting the food into portions. Calculate how many portions the children can make from the following:

   a) 12 steaks to be divided into $\frac{1}{2}$ portions

   b) Four trays of lasagne to be divided into $\frac{1}{10}$ portions

   c) Eight pizzas to be divided into $\frac{1}{6}$ portions

   d) Five cheesecakes to be divided into $\frac{1}{8}$ portions

   e) Six chocolate gateaux to be divided into $\frac{1}{12}$ portions.

2) Draw bar models to solve the following:

   a) $7 \div \frac{1}{3} = ?$      b) $4 \div \frac{1}{5} = ?$      c) $6 \div \frac{3}{4} = ?$

## CHALLENGE!

The children have:

   - 12 bottles of water
   - six bottles of orange juice
   - four bottles of apple juice

They have jugs for the water, large glasses for the orange juice and small glasses for the apple juice.

   - A jug holds $\frac{3}{4}$ of a bottle
   - A large glass holds $\frac{2}{5}$ of a bottle
   - A small glass holds $\frac{1}{3}$ of a bottle

a) How many of each will they need to pour out all of the drinks?

b) How many large glasses could be poured from five bottles of orange juice?

## 6.9 Equivalents: fractions, decimals and percentages

We are learning to convert between fractions, decimals and percentages.

**Before we start**

Convert each of these decimal fractions into a fraction:

0·1     0·25     0·76     2·8

Fractions can be presented in different ways. They have the same value but a different appearance.

**Let's learn**

This bar shows us one half:

| one half | |

we can write this as $\frac{1}{2}$

We can convert **one half** into **tenths**:

   =

one half          five tenths

We can write **tenths** as a **decimal**:   **5 tenths = 0·5**   so   $\frac{1}{2}$ **= 0·5**

We can convert **five tenths** into a **percentage**:

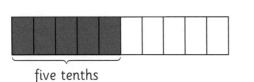   =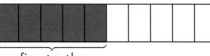

five tenths                    50%

100%

We can write **tenths** as a **percentage**:   **5 tenths = 50%**

So, we can say that:   $\frac{1}{2}$ **= 0·5 = 50%**

**Let's practise**

1) Convert the following fractions into both a decimal and a percentage:

Same value, different appearance!

a)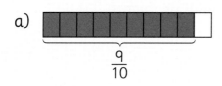

$\frac{9}{10}$

b)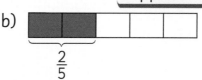

$\frac{2}{5}$

c)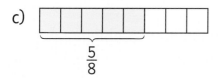

$\frac{5}{8}$

d)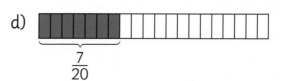

$\frac{7}{20}$

2) Convert each decimal fraction to a percentage and a fraction in its simplest form:

a)  $0 \cdot 6$  =  $\frac{?}{?}$  =  $\frac{?}{?}$  =  ___%

b)  $0 \cdot 85$  =  $\frac{?}{?}$  =  $\frac{?}{?}$  =  ___%

c)  $0 \cdot 34$  =  $\frac{?}{?}$  =  $\frac{?}{?}$  =  ___%

**CHALLENGE!**

Isla has found some fractions that are really tricky to convert to a percentage and decimal fraction:

$\frac{1}{3}$   $\frac{1}{6}$   $\frac{1}{8}$   $\frac{1}{9}$   $\frac{1}{12}$   $\frac{1}{15}$

a) Can you explain why she is finding it difficult to convert these fractions?

b) Convert as many of these fractions as you can for her, rounding off where necessary.

## 6.10 Solving fraction, decimal and percentage problems

We are learning to analyse and solve problems using fractions, decimals and percentages.

**Before we start**

Write each of the following in two other ways:

0·85      $\frac{3}{5}$      24%

Any fraction problem can be presented using fractions, decimals or percentages.

**Let's learn**

There are 32 children in a class:

• Finlay says that $\frac{1}{2}$ of the class walked to school today.

• Nuria says that 0·5 of the class walked to school today.

• Isla says that 50% of the class walked to school today.

The children are giving the same information in different ways. We can solve the problem however we like:

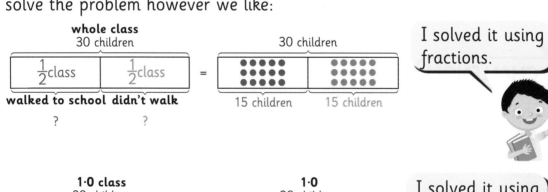

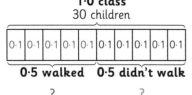

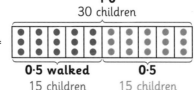

I solved it using fractions.

I solved it using decimals.

 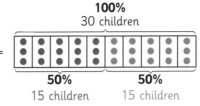 

**100% class**
30 children

| 10% | 10% | 10% | 10% | 10% | 10% | 10% | 10% | 10% | 10% |

**50% walked  50% didn't walk**
?              ?

=

**100%**
30 children

**50%**       **50%**
15 children   15 children

I solved it using percentages.

The answer is the same regardless of the strategy we use.

**Let's practise**

1) Use the bar models to work out the following:

   a) $\frac{2}{3}$ of 5250:

   5250

   ?

   b) 60% of 9500:

   9500

   ?

   c) 0·9 of 7400:

   7400

   ?

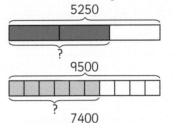

2) Make up a word problem for each of the bar models in question 1.

3) Draw a bar model to solve each of the following problems:

   a) A plane is flying from Glasgow to New York. The total distance is 5160 kilometres. How far is there left to travel if the plane has travelled five-sixths of the journey so far?

   b) 65% of people who viewed a video on YouTube 'liked' the video. How many 'likes' has the video received if it has had 4500 views?

   c) A car's petrol gauge displays 1.0 when the tank is full. The tank holds 75 litres of petrol. How much petrol is left in the tank when the display shows 0·35?

**CHALLENGE!**

Make up and solve a fraction, a decimal and a percentage problem for both of the following bar models:

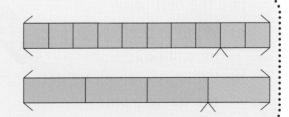

## 7.1 Money problems using the four operations

> We are learning to compare costs using the four operations.

**Before we start**

Isla gave the cashier five £20 notes to pay for her shoes that cost £72·50.

a) How much change will Isla receive?

b) List the possible notes and coins that could be in her change.

**Let's learn**

> We are using our knowledge of decimal notation to compare costs.

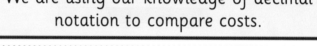

You can buy the same item from different retailers for different prices.

By **comparing prices** you can **save money** on the items that you purchase.

Mr Irwin wanted to buy a new TV and he looked at the same model in three different shops.

In which shop will he receive the best deal?

TV'S "R" US £345

TV'S "R" US £345

ELECTRIC AVENUE £324 **£360 – 10% discount**

ELECTRIC AVENUE £360 with 10% discount

**(£360 ÷ 100 = £3.60, 3·60 × 10 = £36, £360 – £36 = £324)**

PRICE PERFECT £350

PRICE PERFECT £350

*Mr Irwin should buy his TV from Electric Avenue as he will pay the lowest price.*

**Let's practise**

1) Finlay wanted to buy a new computer game so he looked on the internet to find the best price.

Gary's Games – £35·50 + post and packing (£1·25)

Gaming World – £34·75 + post and packing (£2·00)

Gamers Direct – £35·20 free post and packing

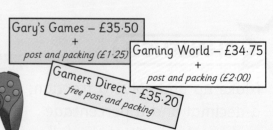

a) What is the lowest price he will pay?

b) Which store will he get the game from?

2) Alex takes the bus every day to and from work except a Friday because he gets a lift.

> **A single ticket costs £1·20**
> **A return ticket costs £2·25**
> **A book of four return tickets costs £8·50**

a) What is the least he will pay for his weekly travel?

b) Which tickets will he buy?

3) Mr Buchanan needed someone to landscape his garden and he had three flyers to choose from:

a) If Mr Buchanan wants someone to work for two days, who should he hire? How much will he pay?

b) If he wants someone for two hours, who should he hire? How much will he pay?

LANDSCAPE LARRY

£5·50 per hour or
£20 per day

GARDENING GURU

£4·00 per hour or
£22·50 per day

GREEN FINGERS
GARDENING SERVICES

£4·50 per hour or
£25 per day

**CHALLENGE!**

Amman wants to buy a DVD player, a TV and DVD boxset.

He compared the prices in three different retailers.

|  | TV'S "R" US | ELECTRIC AVENUE | PRICE PERFECT |
|---|---|---|---|
| DVD | £35·00 | £40·50 | £38·50 |
| TV | £44·50 | £42·75 | £43·00 |
| DVD boxset | £19·90 | £19·95 | £18·75 |

a) Which retailer should he buy each item from to get the best deal?

b) How much does he spend in total?

c) If he bought all three items from ELECTRIC AVENUE, how much would he spend?

## 7.2 Budgeting

We are learning to create a plan to spend within a budget.

**Before we start**

Nuria got £75 in total for her birthday. She wants to buy a new phone that costs £125.

If she gets £5·00 per week pocket money, how many weeks will she need to save before she can afford to buy the phone?

**Budgeting** is a way of creating a plan to spend the money you have.

**Let's learn**

**Budgeting** is one way to **balance** the money you **spend** with the money you **earn**.

Creating a spending plan helps you to plan in advance how much money you will need.

Isla likes puzzle books and buys one every week. She also likes to go skating on a Saturday with her friends. If she gets £30 a month pocket money, she will need to plan how she will spend it.

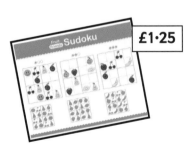
£1·25

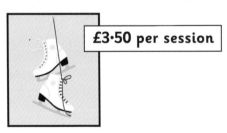
£3·50 per session

If Isla spends £1·25 every week on a puzzle book, how much will she spend per month (*based on a four-week month*)?

$$£1·25 × 4 = £5·00$$

If she spends £3·50 every week on skating, how much will she spend per month?

$$£3·50 × 4 = £14·00$$

To work out how much she will need to keep for these things, the total she spends will need to be added together:

$$£5·00 + £14·00 = £19·00$$

Therefore, Isla needs to budget £19·00 per month for her puzzle books and skating, which leaves her £11·00 (£30·00 − £19·00) to spend on other things.

## Let's practise

1) Look at the following budget plan for Mr Spence and answer the questions:

|  | Credit | Debit |
|---|---|---|
| Income (£250 per week) | £1000 | |
| Rent (£60 per week) | | £240 |
| Food (£40 per week) | | £160 |
| Car (£50 per week) | | £200 |
| Phone (£25 per month) | | £25 |
| | | |
| Totals | £1000 | £625 |

  a) How much does Mr Spence have left once he has paid his bills?
  b) If he wants to get a gym membership that costs £22 per week, how much will he have left every month.
  c) Can he afford to subscribe to satellite TV that will cost £55 per month?

2) If Finlay gets £10 a week pocket money and £15 every month from his grandparents, copy and complete the following spending plan:

| | Credit | Debit |
|---|---|---|
| Income | £ | |
| Comic (£1·50 per week) | | £ |
| Cinema (£5·00 per week) | | £ |
| | | |
| Totals | £ | £ |

3) Mrs Roberts earns £1250 per month, her mortgage is £500 per month and she spends £50 per week for food. How much does she have left to spend each month?

## CHALLENGE!

Using the following information, create a budget for the Irwin family:

| | |
|---|---|
| Mrs Irwin's monthly income | £1000 |
| Mr Irwin's monthly income | £1000 |
| Mortgage per month | £650 |
| Food per month | £400 |
| Electricity and gas per month | £75 |
| Mrs Irwin's phone per month | £30 |
| Mr Irwin's phone per month | £35 |
| Car per month | £150 |
| Family gym membership per month | £50 |
| Satellite TV subscription per month | £45 |

a) How much do the Irwin family have left to spend at the end of every month once they have paid everything?

b) If they want to save £200 per month how much will they have left to spend?

## 7.3 Profit and loss

We understand and apply the terms profit and loss.

### Before we start

1) Mr Jones sold a painting for a profit of £175. If the selling price was £725, how much did he buy the painting for?

2) Mrs Lawrence is selling five cakes that cost her £2·50 each to make.

    If she wants to make a total profit of £10, how much will she need to sell each cake for?

To calculate the **profit**, you need to deduct the selling price from the original purchase price.

### Let's learn

In order to make a **profit** on something, you need to sell the item for **more than** you purchased it for.

| | |
|---|---|
| **Purchase price** | £250 |
| **Selling price** | £325 |
| **Profit** | **£75** |

You will make a **loss** on something if you sell it for **less than** you purchased it for.

| | |
|---|---|
| **Purchase price** | £740 |
| **Selling price** | £690 |
| **Loss** | **£50** |

## Let's practise

1) Amman bought a phone that cost £225, a case that cost £12 and headphones that cost £25. If he sold them all for a total of £232, did he make a profit or loss and if so by how much?

2) Isla's dad bought a car for £8130 and sold it six months later for a profit of £755. How much did he sell it for?

3) A shopkeeper bought 40 Easter eggs for a total cost of £70.

He sold 30 of them for £3 each before Easter and the remaining 10 after for £1 each.
What was the total profit that the shopkeeper made?

## CHALLENGE!

Finlay and his classmates want to raise £350 for charity.

They have decided to put on a play in order to raise the money.

It will cost them £50 to hire the stage, £100 to buy the costumes and £25 to print the posters, tickets and programmes (unlimited amount).

The hall can hold 50 people in the audience.

How many tickets and at what price will they need to sell in order to cover their costs and make the £350 profit?

# 7 Money

## 7.4 Discounts

We are learning to investigate different pricing structures.

**Before we start**

Lauren wants to buy a new computer and has found three different deals.

She gets a 10% student discount in all stores in addition to any offers that they have – *therefore, she would get a 25% discount at TECH GEEKS...*

TECH GEEKS- £350 plus 15% discount

COMPUTECH- £345 plus £30 off

GADGET GARAGE- £330 with extra 20% student discount

1) Where should Lauren buy her computer from?

2) What is the least Lauren will pay for her computer?

**Discounts**, **deals** and **loyalty cards** offered by retailers allow us to save money when we purchase things.

**Let's learn**

Many retailers use **discounts** and different **deals** to encourage customers to buy their products.

If you compare all the different discounts and deals that each retailer offers you can work out where you can buy the product you want and save the most amount of money.

Many retailers also have their own **loyalty cards** that offer additional benefits to the customer.

LOYALTY CLUB

*customer appreciation card*

*Loyalty cards offer customers a variety of different benefits:*

*Points – One point for every pound spent*

*Discount – additional 10% off*

*Money off – £1 off for every £10 spent*

**Let's practise**

1) Finlay wants to buy a new computer game that costs £75.

   He has a loyalty card for the store that will give him £1 off for every £10 he spends.

   The store also has 20% off all games.

   How much will he **pay** for the game?

2) Isla wants to buy new earrings that cost £16 per pair and she wants two pairs.

   She has a loyalty card for the store with an offer of 'buy one get one half price' for earrings. The shop also has 10% off all items.

   She can choose only one offer, 'buy one get one half price' or 10% off.

   Which offer will **save** Isla the most money?

**CHALLENGE!**

| FOOD & STUFF |
| --- |
| Loyalty Card – £1 off for every £10 spent |
| Discounts – 10% off all fruit and veg |
| Offers – buy one get one half price on all tinned items |

| SUPERMARKET |
| --- |
| Loyalty Card – £1 off for every £20 spent |
| Discounts – 15% off all fruit and veg |
| Offers – buy one get one free on all tinned items |

The prices at each supermarket are the same, so look at the shopping list and decide where Nuria should buy the items she needs:

| **Shopping List** |
| --- |
| 4 bananas |
| 6 apples |
| 5 oranges |
| 1 watermelon |
| 4 tins of tomatoes |
| 2 tins of mushroom soup |
| 2 tins of beans |

| | |
| --- | --- |
| Bananas | 50p each |
| Apples | £2·00 for six |
| Oranges | 70p each |
| Watermelon | £1·50 each |
| One tin of tomatoes | 70p |
| One tin of mushroom soup | 80p |
| One tin of beans | 50p |

What is the least that Nuria can spend?

## 7.5 Hire purchase

We are learning about hire purchase.

**Before we start**

Discuss with a partner the following banking terms and write down what they are:

- overdraft
- direct debit
- standing order

You can use **hire purchase** to pay for something in instalments over a period of time.

**Let's learn**

A **hire purchase** (or **HP**) agreement is a way to buy something by paying a small **deposit** and agreeing to pay the **balance** on a regular basis until the total amount is paid.

Mr Roberts bought a new fridge freezer.

He decided to take out a hire purchase agreement with the retailer.

£260

He paid a **deposit** of £50 and agreed to pay 12 **monthly payments** of £20.

Deposit = £50

Monthly payments = £240 (12 × £20)

Total = **£290***

**\*CAN YOU SEE THAT YOU PAY MORE FOR THE FRIDGE FREEZER USING HIRE PURCHASE?**

**Let's practise**

1) Discuss with a partner why people would use hire purchase to buy goods. Write down the pros and cons of using this method of payment.

2) Mrs Lewis wanted to buy a new washing machine that costs £220.

She can buy it using a hire purchase agreement.

a) Copy and complete the following:

Deposit = £30
Monthly payments = ___? (12 × £20)
Total HP cost = ?

b) How much more will Mrs Lewis pay by using this payment method?

3) Nuria's dad bought a bike for £175 from CYCLE RIGHT.

There was no deposit required and no extra cost by choosing to pay using a Hire Purchase agreement.

a) If he paid it back over seven months, how much did he pay each month?

b) If he paid a £15 deposit and paid it back over four months, how much did he pay each month?

## CHALLENGE!

Isla's family wanted to book a holiday.

They looked at three different companies who offered different rates of hire purchase for a £2500 holiday for a family of four.

1) What is the total price each company have charged for each holiday?

2) Which company is the most expensive?

| | Deposit | Payments |
|---|---|---|
| **Holiday Heaven** | 10% | 10 monthly payments of £230 |
| **Sunny Seas** | £500 | 12 monthly payments of £180 |
| **Beach Bound** | £50 per person | 10 monthly payments of £250 |

# 8 Time

## 8.1 Investigating how long a journey will take

We are learning to investigate how long a journey will take.

**Before we start**

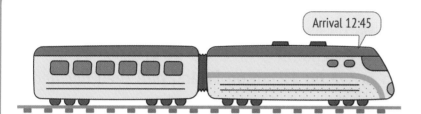

Arrival 12:45

Isla was travelling by train to visit her cousin.

The train left the station at 11:30 am and arrived at the station at 12:45 pm.

How long was Isla's journey?

The **total length** of a journey can be calculated by looking at the **time interval** between the **start** and the **end** of the journey.

**Let's learn**

The **total length** of a journey is calculated by looking at the **time interval** or the **elapsed time** between the **start** of the journey and the **end** of the journey.

It is important to be able to calculate this information **accurately** so that you can plan your journey in advance.

Start time – <u>3·35 pm</u>     End time – <u>5·10 pm</u>

25 minutes     1 hour     10 minutes

3·35 pm     4·00     5·00     5·10 pm

elapsed time = <u>1 hour and 35 minutes</u>

1) A group of fundraisers decided to travel from Stirling to Aviemore, which is approximately 116 miles. They used a satellite navigation system to find out some information:

- It is a 2 hour 29 minute drive.
- It would take 1 day 12 hours to walk.
- It is a 2 hour 58 minute bus ride.

   a) If they decided to go by car and left Stirling at 9 am, what time would they arrive in Aviemore?
   b) If they wanted to arrive by 12 pm on Friday and they decided to walk, when would they have to leave Stirling?
   c) If the bus left Stirling station at 2 pm, what time would they arrive in Aviemore?

2) If Finlay's train leaves the station at 10·45 am and arrives at his destination 3 hours and 20 minutes later, what time did he arrive?

3) The flight to London leaves Edinburgh Airport at 10·25 am and arrives at London Heathrow at 11·32 am.

If Mr Irwin leaves his house at 10 am and arrives at his hotel at 12 pm...
   a) How long is Mr Irwin's journey in total?
   b) How long was Mr Irwin's flight?
   c) How long did it take Mr Irwin to travel from London Heathrow to his hotel?

**CHALLENGE!**

Work with a partner to find out the person in your class who has the longest journey to school (total time travelling) by collecting the following information:

- What time do they leave the house?
- What time do they arrive at school?
- Do they stop anywhere on the way?

# 8 Time

## 8.2 Calculating the duration of activities and events

> We are learning to calculate the duration of activities and events.

### Before we start

Write down the following information:

- The time your school day starts and ends.
- The time your morning break starts and ends.
- The time your lunch break starts and ends.

Now, answer the following questions:

a) How long is your school day?

b) What is the duration of your morning break?

c) How much time do you have for your lunch break?

> The **duration** is the length of time that an event lasts.

### Let's learn

We measure the **duration** of an event in *hours*, *minutes* and *seconds*.

With units of time we can calculate how long an event or activity has taken or will take.

With units of time we can compare events, deciding which took place over a longer or shorter period of time.

1) Work with a partner. Each choose a counter and place it on the 'start' line. Take turns to spin a spinner (1–2) and move that number of spaces along the track. If you land on a task, estimate how long you think it will take. Then complete it while your partner times you with a stopwatch. If your completed time is within 30 seconds of your estimate, take an extra turn.

| | Draw and cut out a rectangle 10 cm by 12 cm | | Count backwards in 5s from 200 to 0 | | **Start** |
|---|---|---|---|---|---|
| Write 10 five-letter words | | | | | |
| | Find five things in the classroom that begin with the letter 'b' | | Add together the ages of all the students in your class | | Walk the length of your classroom 30 times |
| | | | | | |
| **Finish** | Find and weigh a set of objects that total almost 1 kg | | Throw and catch a ball 40 times with a partner | | Write the names of all the students in your class in alphabetical order |

2) Calculate the duration of each event using relevant time units (seconds, minutes, hours or days):

    i) Nuria is solving a maths problem. The stopwatch shows 00:00:00 (Hours: Minutes: Seconds) when she begins and 00:03:00 when she completes it. How long did she take?

    ii) A writing competition begins at 4:00 pm on Tuesday and ends at 4:00 pm on Thursday. What is the duration of the competition?

3) Calculate the duration of each event using relevant time units (seconds, minutes, hours or days).

    i) A man begins work at half past eight in the morning and finishes at five o'clock. How long is he at work?

    ii) Amman's family leave for their holiday on 2nd March and return on 10th March. How long are they away?

**CHALLENGE!**

Calculate the duration of each event using relevant time units:

    i) A digital clock reads 14:15:23 (Hours: Minutes: Seconds) when Saeed begins his walk and 18:24:17 when he finishes. How long is his walk?

    ii) Isla visits a friend. Her digital watch reads 15:07 (0 seconds) when she leaves home on 3rd June and reads 13:05 when she returns home on 5th June. How long has Isla been away from home?

## 8.3 Investigating ways speed, time and distance can be measured

> We are learning to investigate ways that Speed, Distance and Time can be measured.

**Before we start**

Using the formula **distance = speed × time**, calculate the distance travelled by a:

a) train, travelling at 80 mph for four hours

b) police car, doing 70 mph for two hours

c) coach, going at 50 mph for five hours.

> I can use the speed, time and distance formulas to solve problems.

**Let's learn**

To find the speed, distance is over time in the triangle below, so speed is distance divided by time.

### SPEED = DISTANCE ÷ TIME

To find time, distance is over speed, so time is distance divided by speed.

### TIME = DISTANCE ÷ SPEED

To find distance, speed is beside time, so distance is speed multiplied by time.

### DISTANCE = SPEED × TIME

These formulas can be summarised in this diagram:

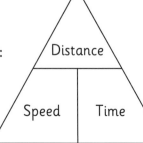

**Let's practise**

1) Use the formula **time = distance ÷ speed** to calculate the time taken for each of these journeys:
   a) Isla walked 10 miles at 2 mph.
   b) Finlay ran 16 km at 4 kph.
   c) Mr Smith drove 500 miles at 50 mph.
   d) Lauren cycled 36 km at 9 kph.

2) Use the formula, **speed = distance ÷ time**, to calculate the average speed of these journeys:
   a) A train travelling 480 miles in six hours.
   b) A plane flying 2700 kilometres in 10 hours.
   c) A car driving 560 miles in eight hours.
   d) A man running 12 miles in three hours.

3) Use the correct formula to answer the following questions:
   a) Find the distance travelled by a car driving at 45 mph for five hours.
   b) Find the average speed that a train travelled at, covering a distance of 100 miles for two hours.
   c) Find the time taken by a cyclist travelling at a speed of 30 kph over 150 km.

**CHALLENGE!**

Isla lives five minutes away from the school.

The distance from her house to the school is 800 metres.

a) If Isla walks to school, what will her average speed be?

b) If she cycles, she will get there in half the time, how long will her bike journey be?

# 8 Time

## 8.4 Calculating time accurately using a stopwatch

We are learning to use a stopwatch to calculate time.

### Before we start

Answer true or false to the following statements:
a) Time can be measured using a calendar.
b) A millimetre is a unit of time.
c) You can use a ruler to measure time.
d) A week is equal to 192 hours.

A stopwatch is used to calculate the time that has elapsed from the start to the end of an activity.

### Let's learn

A stopwatch will calculate the time an activity takes using minutes, seconds and centiseconds (hundredths of a second).

You would use a stopwatch to record short activities and sporting events that require accurate and precise timing.

### Let's practise

1) Work with a partner. Take turns to perform these tasks while your partner uses a stopwatch to measure the duration of the task. When you have completed the task, say the time on the stopwatch to your partner.

| Write a list of animals beginning with the letters A to J | |
| Count backwards from 100 to 0 in 2s | |
| Draw and colour a rainbow | |

2) Estimate the amount of time each task in the table will take. Complete each task and record the duration of the event in minutes and seconds. How close are your estimates?

| Task | Estimate | Duration |
|---|---|---|
| Jump up and down on the spot 50 times | | |
| Build a tower of cubes 30 cubes tall | | |

3) Work with a partner. Complete the tasks and record the duration of each event in minutes and seconds. Then answer the questions.

| Task | You | Your partner |
|---|---|---|
| Write each multiple of 6 from 6 to 180 | | |
| Construct a cuboid 5 × 6 × 7 from interlocking cubes | | |
| Fold a piece of paper in half six times | | |

Which task took you longest?

Which task took your partner the shortest amount of time?

CHALLENGE!

Complete the table by calculating the amount of time between start and finish times.

| Start time | Finish time | Time elapsed |
|---|---|---|
| 5 minutes 37 seconds | 9 minutes 49 seconds | |
| 3 minutes 54 seconds | 11 minutes 37 seconds | |
| 11 minutes 57 seconds | 23 minutes 8 seconds | |
| 28 minutes 59 seconds | 57 minutes 13 seconds | |

# 8 Time

## 8.5 Converting between units of time

We are learning to convert between units of time.

### Before we start

For each of the following tasks, use the correct unit of time to describe how long the activity might take (second, minute, hour, day, week, month, year, century):

a) bake a cake
b) fly to Australia
c) hop
d) an oak tree to grow

A unit of time is any time interval that is used as a way of measuring duration.

### Let's learn

The basic units of time are **second**, **minute**, **hour**, **day**, **week**, **month**, **year**, **century** and **millennium**.

A **second** is one of the **smallest** units of time.

Can you estimate how long a second is?

If you raise your hand up and immediately bring it back down, this will have taken you about a second!

**Let's practise**

1) Match the following units of time:

| | |
|---|---|
| Fortnight | Decade |
| 48 hours | 100 years |
| 10 years | two weeks |
| Century | two days |

2) Convert the following from seconds to minutes:
   a) 300 seconds   b) 540 seconds   c) 900 seconds   d) 1800 seconds

3) Convert the following from minutes to hours and minutes:
   a) 240 minutes   b) 510 minutes   c) 390 minutes   d) 750 minutes

4) Convert the following from hours to days:
   a) 96 hours      b) 168 hours     c) 336 hours     d) 720 hours

**CHALLENGE!**

Investigate with a partner the following units of time:
millisecond, microsecond, decasecond, quarter, olympiad, jubilee and mega-annum.
Can you convert them to more well-known units of time?

# 9 Measurement

## 9.1 Estimating and measuring length

We are learning to estimate and measure lengths to two decimal places.

### Before we start

Convert the following measurements into centimetres:

68 mm          1·3 m          7 mm          17·5 m

The more accurately we measure, the more decimal places our measurement will be given to.

### Let's learn

Length can be measured more accurately when we use decimals. Rather than saying the distance from home to school is between one and two kilometres, we can say it is 1·4 km:

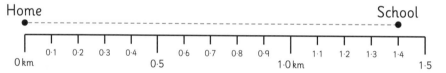

We can partition this value to make more sense of it:

| 1·4 km | = | 1 km | + | 0·4 km |
|--------|---|------|---|--------|
|        | = | 1000 m | + | 400 m |

**1·4 km    =    1400 m**

> The first decimal place represents tenths of a kilometre.

We can be even more accurate if we zoom in on the measurement and introduce the second decimal place:

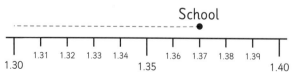

| 1·37 km | = | 1 km | + | 0·3 km | + | 0·07 km |
|---------|---|------|---|--------|---|---------|
|         | = | 1000 m | + | 300 m | + | 70 m |

**1·37 km  =  1370 m**

> The second decimal place represents hundredths of a kilometre.

We can now say, more accurately, that the distance from home to school is exactly 1·37 km or 1370 m.

It is essential to measure accurately when designing or building something in order that everything fits together properly.

### Let's practise

1) An architect has given measurements in centimetres but the builder has asked for them to be converted into metres:

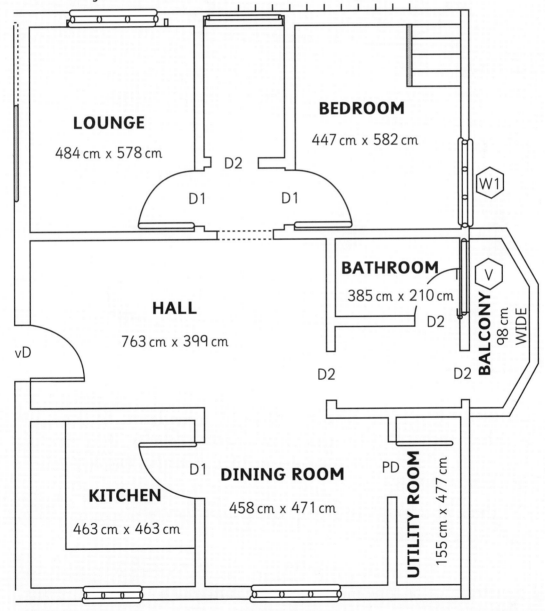

Convert each of the measurements into metres.

2) A racing car team need accurate measurements for each circuit to work out exactly how much fuel to put in the tank for each race. Convert each of the following into kilometres for them:

a)

Australia: 5310 m

b)

Spain: 4650 m

c)
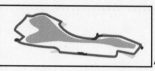
Britain: 5890 m

d)

Germany: 4570 m

3) The following insect and arachnid lengths have been given in millimetres. Convert each measurement into centimetres:

a) Beetle = 22·4 mm
b) Housefly = 9·7 mm
c) Spider = 13·8 mm
d) Cockroach = 30·5 mm
e) Crane fly = 42·0 mm
f) Scorpion = 93·1 mm

**CHALLENGE!**

a) Choose six items from your environment around you. Estimate the length of each item in metres to two decimal places, then measure each item accurately with a ruler or tape measure. Display your results in a suitable table.

b) Use chalk or a length of string. Draw lines (or cut pieces of string) that you estimate to be:

- 37 cm
- 125 cm
- 238 cm

Use a metre stick or tape measure to measure the lengths accurately. Convert your measurements to metres.

# 9 Measurement

## 9.2 Estimating and measuring mass

We are learning to estimate and measure the mass of an object to two decimal places.

**Before we start**

Write the following weights in grams and kilograms:

The more accurately we measure mass, the more decimal places the measurement will be given to.

**Let's learn**

1 kg = 1000 g

$\frac{1}{100}$ kg = 10 g

One kilogram can be split into 100 equal parts of 10 grams. 10 grams is one hundredth of a kilogram. We can write this as a decimal:

**10 g = $\frac{1}{100}$ kg = 0·01 kg**

The weight of this bag is between 1·5 and 1·6 kilograms:

We can be more accurate if we zoom in on the scale:

We can now say, more accurately, that the weight of the bag is exactly 1·53 kg or 1530 g.

The weight of the bag hasn't changed. We have simply measured it with greater accuracy.

## Let's practise

1) Nuria is packing a suitcase for her holiday. She weighs the contents to check she's not over her allowance for the plane.

a) Write the mass of her contents in both grams and kilograms:

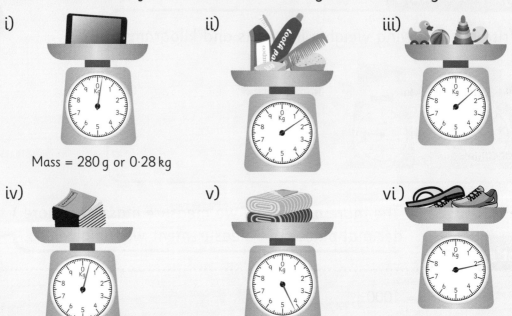

i)

Mass = 280 g or 0·28 kg

ii)

iii)

iv)

v)

vi)

b) The weight allowance for the plane is 10 kg. Advise Nuria about her trip.

2) Identify two items in the environment around you that you estimate to have a mass of between:

a) 0 kg and $\frac{1}{2}$ kg        b) $\frac{1}{2}$ kg and 1 kg        c) 1 kg and $2\frac{1}{2}$ kg

Display your results in a suitable table.

### CHALLENGE!

You will need suitable scales and a material to be weighed such as sand, and a bag or container. Fill the bag or container with the materials and measure out the following weights:

- 0·05 kg
- 0·14 kg
- 0·56 kg
- 1·04 kg

# 9.3 Estimating and measuring area

**Before we start**

Calculate the area of the following shapes:

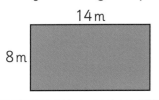

14 m

8 m

13 m

15 m

We are learning to estimate and measure area to two decimal places.

Partitioning helps us work out the area of a rectangle.

**Let's learn**

Finlay isn't sure how to work out the area of this rectangle:

Isla suggests partitioning it into two smaller rectangles and adding the two areas together:

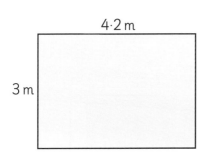

4·2 m

3 m

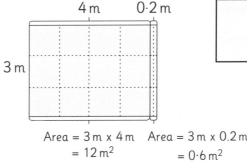

4 m    0·2 m

3 m

Area = 3 m x 4 m
= 12 m²

Area = 3 m x 0.2 m
= 0·6 m²

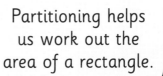

1 square metre

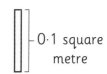

0·1 square metre

We can calculate the area of the larger part by multiplying 3 m by 4 m to get 12 m².

The smaller part measures 3 m by 0·2 m (or two tenths of a metre).

3 multiplied by two tenths is six tenths, therefore, 3 m x 0·2 m = 0·6 m².

The total area of the rectangle can be worked out by adding the areas of the two parts together:

**Area = 12 m² + 0·6 m² = 12·6 m²**

I estimate that it will be just over 12 m² because 3 × 4 = 12.

# 9

1) Copy the following rectangles and calculate their areas (show how they can be partitioned):

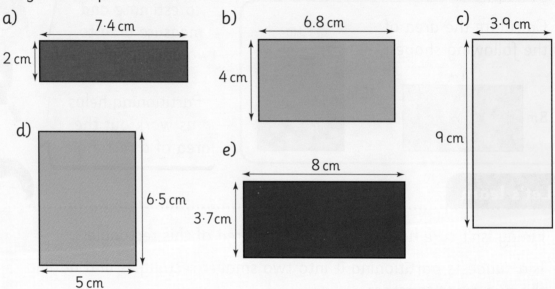

a)
7·4 cm
2 cm

b)
6.8 cm
4 cm

c)
3·9 cm
9 cm

d)
6·5 cm
5 cm

e)
8 cm
3·7 cm

2) Find the missing lengths for each of the following surface areas:

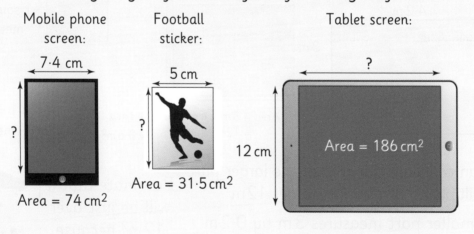

Mobile phone screen:
7·4 cm
?
Area = 74 cm²

Football sticker:
5 cm
?
Area = 31·5 cm²

Tablet screen:
?
12 cm
Area = 186 cm²

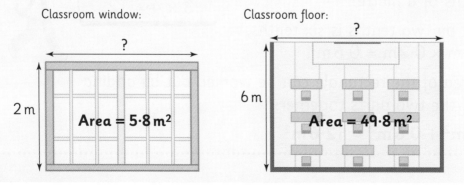

Classroom window:
?
2 m
Area = 5·8 m²

Classroom floor:
?
6 m
Area = 49·8 m²

CHALLENGE!

Finlay thinks he's worked out the area of this rectangle without a calculator:

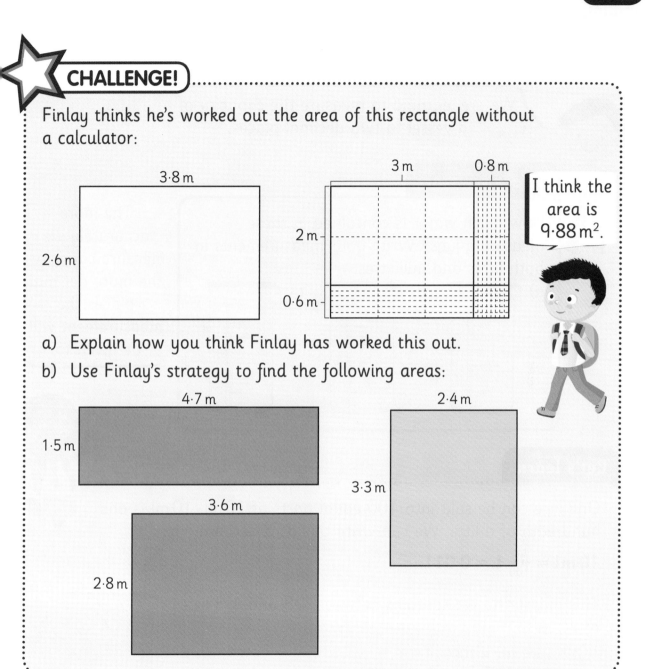

3.8 m

2.6 m

3 m    0.8 m

2 m

0.6 m

I think the area is 9·88 m².

a) Explain how you think Finlay has worked this out.

b) Use Finlay's strategy to find the following areas:

4.7 m

1.5 m

2.4 m

3.3 m

3.6 m

2.8 m

## 9.4 Estimating and measuring capacity

We are learning to measure the capacity of a vessel to two decimal places.

### Before we start

How much water is contained in these measuring jugs? Write your measurements in both litres and millilitres.

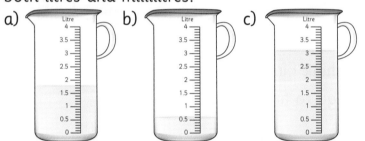

a) b) c)

The more accurately we measure capacity, the more decimal places the measurement will be given to.

### Let's learn

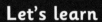

One litre can be split into 100 equal parts of 10 ml. 10 ml is one hundredth of a litre. We can write this as a decimal:

$$10\,ml = \frac{1}{100}\,L = 0\cdot01\,L$$

This measuring jug contains between 1·8 and 1·9 litres of water:

1 litre = 1000 ml

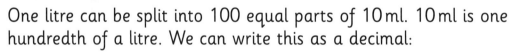

$\frac{1}{100}$ litre = 10 ml

We can be more accurate if we zoom in on the scale:

We can now say, more accurately, that the measuring jug is filled to exactly 1·86 L or 1860 ml.

The amount of water hasn't changed. We have simply measured it with greater accuracy.

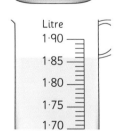

## Let's practise

1) Isla is making a fruit punch that contains the following:

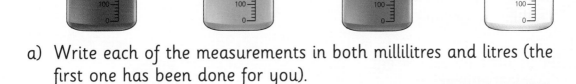

i) pineapple juice

millilitres: 140 ml; litres: 0.14 L

ii) cranberry juice

iii) mango juice

iv) blackcurrant juice   v) orange juice   vi) apple juice   vii) soda water

a) Write each of the measurements in both millilitres and litres (the first one has been done for you).

b) Calculate the total volume of the fruit punch in millilitres and litres.

2) Use a measuring jug to measure out the following volumes of water, then write each measurement in litres (the first one has been done for you):

a) 120 ml = 0·12 L      b) 70 ml = _____

c) 160 ml = _____      d) 330 ml = _____

e) 850 ml = _____      f) 1240 ml = _____

You may need to fill more than one measuring jug for some of these volumes of water!

**CHALLENGE!**

You will need:

- One full two-litre bottle of water
- One empty two-litre bottle of water
- A funnel
- A measuring jug
- To work with a partner.

Repeat the instructions below for each of the following measurements:

a) 0·04 L    b) 0·25 L    c) 0·38 L    d) 0·73 L    e) 1·16 L    f) 1·67 L

i) Using estimation, pour water into the empty bottle. (Use the funnel while your partner holds the bottle steady.)

ii) Empty the water into the measuring jug.

iii) Read the measurement.

This measures 0·6 L or 600 ml.

iv) Copy and complete the table:

|  | Actual (litres) | Actual (millilitres) |
|---|---|---|
| a) |  |  |
| b) |  |  |
| c) |  |  |
| d) |  |  |
| e) |  |  |
| f) |  |  |

Calculate how close you were for each attempt.

# 9 Measurement

## 9.5 Estimating imperial measurements

> We are learning to estimate imperial measurements by comparing familiar objects.

**Before we start**

Estimate the following:
- The height of your teacher's desk in centimetres.
- The weight of your school bag in grams or kilograms.
- The capacity of your (or a friend's) water bottle in millilitres or litres.

Share your answers with a partner and check your measurements.

> We can use the size for something we know to estimate the measurements of other things.

**Let's learn**

| Length | Mass | Capacity |
|---|---|---|
| 1 foot = 12 inches | 1 pound = 16 ounces | 1 cup = 10 fluid ounces |
| 1 yard = 3 feet | 1 stone = 14 pounds | 1 pint = 2 cups |
| 1 mile = 1760 yards | 1 tonne = 160 stone | 1 gallon = 8 pints |

> I am four feet tall. I estimate that my teacher is six feet tall.

> I weigh 80 pounds. I estimate that my dad weighs 160 pounds.

> The glass holds one pint of water. I estimate that the jug holds three pints of water.

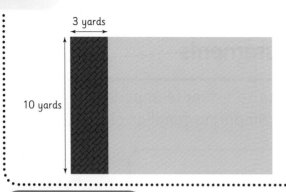

3 yards

10 yards

The patio is 30 square yards. I estimate that the lawn is 120 square yards.

## Let's practise

1) Mount Snowdon in Wales is 3560 feet high. Compare this to each of the other mountains below to estimate their height. Can you estimate their heights within 500 feet?

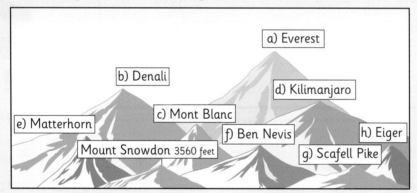

a) Everest
b) Denali
d) Kilimanjaro
e) Matterhorn
c) Mont Blanc
f) Ben Nevis
h) Eiger
g) Scafell Pike
Mount Snowdon 3560 feet

2) The distance from Glasgow to Edinburgh is 42 miles. Estimate the distance for each of the following:

a) Glasgow to Ayr
b) Dundee to Inverness
c) Edinburgh to Aberdeen
d) Edinburgh to Perth
e) Dumfries to Inverness
f) Perth to Dundee

CHALLENGE!

The Ice Hotel in Sweden is made entirely out of blocks of ice.

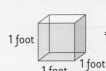 = $3\frac{1}{2}$ stone    i) 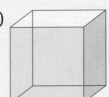    ii)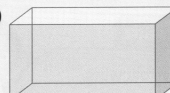

iii) 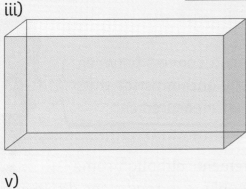    iv)

v)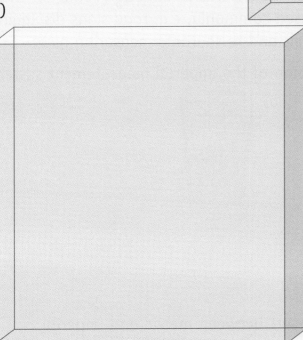

Each block is exactly one foot thick.

A cubic foot of ice weighs approximately $3\frac{1}{2}$ stone.

a) Estimate the weight of each block of ice.

b) Draw designs for the following items in the hotel and estimate their weights:
   • bench   • table   • bed

## 9.6 Converting imperial measurements

**Before we start**

Convert the following:

- $4\frac{1}{2}$ feet to inches

- 5 pounds to ounces

- 20 pints to gallons

We are learning to convert between metric and imperial measurements.

We can convert between metric and imperial units of measurement.

**Let's learn**

The UK uses the metric system for measurement, although some imperial measurements are still used commonly, e.g. road signs displaying miles and drinking glasses in hotels and restaurants measuring pints.

The USA still uses their own version of the imperial measurement system.

a)

| A 87 | |
| --- | --- |
| Coal Acain Kyleakin | 2 Miles |
| An t-Ath Leathann Broadford | 9 Miles |
| Port Righ Portree | 34 Miles |
| Uige Uig | 49 Miles |

b)

1 pint

We can use the following to convert between metric and imperial measurements:

**Length**

1 inch = 2·54 cm
1 foot = 30·5 cm
1 yard = 91·4 cm
1 mile = 1·61 km

**Mass**

1 ounce = 28·3 g
1 kg = 2·2 pounds
1 stone = 6·35 kg

**Capacity**

1 fluid ounce = 28·4 ml
1 cup = 240 ml
1 litre = 1·76 pints
1 gallon = 4·55 litres

The scales say I weigh 63 lb (pounds). How do I calculate my weight in kilograms?

63 lb ÷ 2·2 = **28·6 kg**     Divide your weight by 2·2.

I am 53 inches tall. How do I calculate my height in centimetres?

53 inches × 2·54 = **134·6 cm**

Multiply your height by 2·54.

**Let's practise**

1) Convert the following:

   a) 15 cm to inches
   b) $2\frac{1}{2}$ m to feet
   c) 75 g to ounces
   d) 3·6 kg to pounds
   e) 50 kg to stone
   f) 12 fluid ounces to millilitres
   g) 25 pints to litres
   h) 10 litres to gallons

2) The measurements for a football penalty area are normally given in yards:

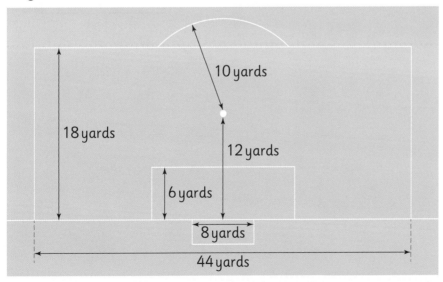

10 yards

18 yards

12 yards

6 yards

8 yards

44 yards

Convert each of the measurements to both metres and centimetres.

3) The ingredients for a trifle are given in imperial measurements.
Convert the ingredients to metric measurements:

$1\frac{1}{2}$ pounds summer fruits

8 ounces caster sugar

$2\frac{1}{2}$ ounces custard powder

2 pints milk

$\frac{1}{4}$ pound madeira cakes

$\frac{1}{2}$ cup sherry

20 fluid ounces double cream

$\frac{1}{2}$ fluid ounce vanilla extract

**CHALLENGE!**

The dimensions for an aquarium are as shown:

Calculate the capacity of the aquarium in:

a) litres

b) gallons

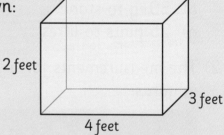

2 feet

3 feet

4 feet

## 9.7 Calculating perimeter (rectilinear shapes)

We are learning to calculate the perimeter of rectilinear shapes.

**Before we start**

Draw a shape that has a perimeter of:
- 10 cm
- 14 cm
- 15 cm

We can work out the **perimeter** of any shape by adding up the lengths of all the sides.

**Let's learn**

A **rectilinear shape** is any shape all of whose edges meet at right angles.

Calculating perimeter is really important in everyday life. It's useful if we want to:

- Work out how much fence to buy to put around a garden.
- Draw plans to build a house, barn or a playground.
- Calculate how far it is to run around the school playing fields, etc.

**Perimeter** is the total distance around the outside of any shape. It is calculated by adding up the lengths of all of the sides.

The children have been asked to help plant trees around the perimeter of the school playground. They have to plant trees one metre apart.

To work out how many trees we will need, we will have to measure the perimeter first:
4 + 3 + 3 + 3 + 7 + 6 = **26 m**

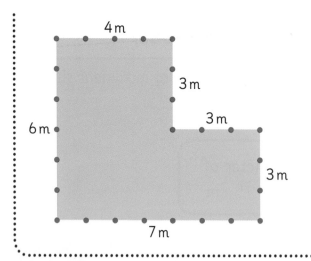

4 m

3 m

3 m

6 m

3 m

7 m

The perimeter of the
playground is 26 m so they
will need to plant 26 trees.

## Let's practise

1) Calculate the perimeter of each of the shapes below:

a)

4 cm

2 cm

3 cm

5 cm

3 cm

7 cm

b)

6 cm

3 cm

4 cm

3 cm

c)

2 cm  5 cm  3 cm

4 cm

10 cm

d)

5 cm

square

square  3 cm

5 cm

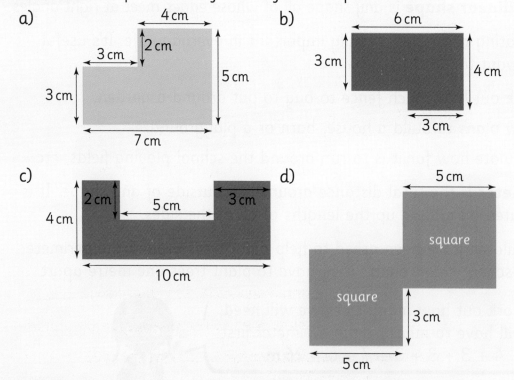

2) The children have entered a competition to design a new layout
for the school playground. Calculate the perimeter of each section
of their design:

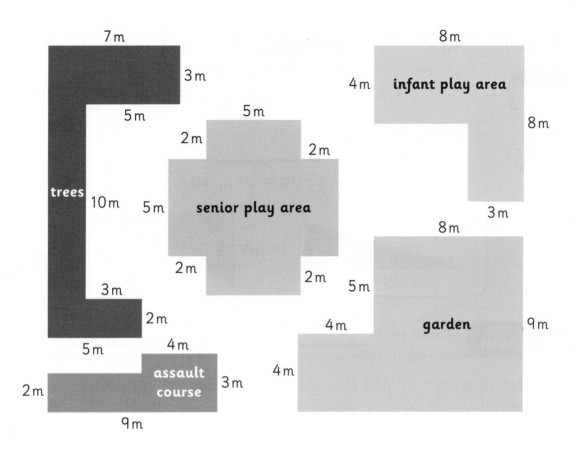

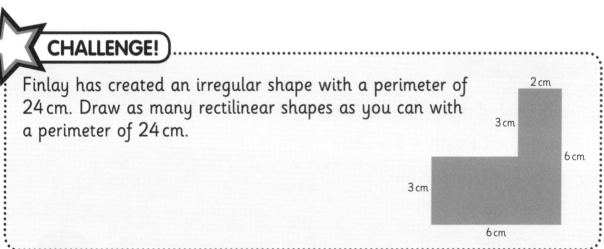

**CHALLENGE!**

Finlay has created an irregular shape with a perimeter of 24 cm. Draw as many rectilinear shapes as you can with a perimeter of 24 cm.

## 9.8 Calculating area: triangles

 We are learning to calculate the area of any triangle.

**Before we start**

Calculate the area of these triangles:

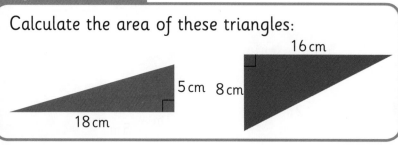

16 cm

5 cm  8 cm

18 cm

Calculate the area of these triangles:

**Let's learn**

The area of a right-angled triangle can be calculated by drawing imaginary lines to create a rectangle:

6 cm

8 cm

The area of this triangle is exactly half the area of the rectangle.

We can calculate the area of a right-angled triangle using the formula:

**Area = $\frac{1}{2}$ × Length × Breadth:**

Area = $\frac{1}{2}$ of 8 cm × 6 cm

$= \frac{1}{2}$ × 48 cm$^2$

$= 24$ cm$^2$

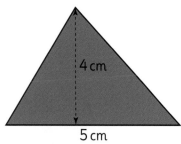

4 cm

5 cm

How can we calculate the area of this triangle?

We can draw two imaginary triangles to create a rectangle.

The original triangle is exactly half the size of the rectangle so we can use the same formula:

**Area = $\frac{1}{2}$ × length × breadth**

$= \frac{1}{2}$ × 5 cm × 4 cm

$= \frac{1}{2}$ × 20 cm²

**= 10 cm²**

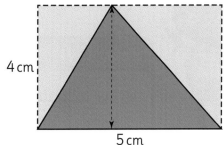

**Let's practise**

1) Calculate the area of each of the following:

a)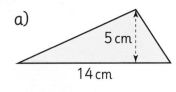
5 cm
14 cm

b)
8 cm   8 cm

c)
12 cm
9 cm

2) Using squared paper, draw triangles that have the following areas (you may find it helpful to start by drawing a rectangle):

a) 50 cm²   b) 21 cm²   c) 36 cm²   d) 60 cm²   e) 39 cm²

⭐ **CHALLENGE!**

Finlay has been challenged to calculate the area of the following two shapes:

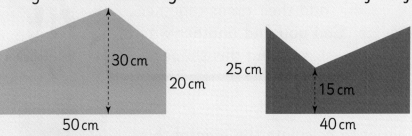

30 cm   20 cm   50 cm

25 cm   15 cm   40 cm

i)  Explain how Finlay could solve this problem.

ii)  Help Finlay complete his challenge.

## 9.9 Measuring area (composite shapes)

We are learning to calculate the area of composite shapes.

**Before we start**

Draw a square, rectangle or triangle with an area of:
- 6 cm²
- 9 cm²
- 12 cm²

Shapes can be split into smaller parts to help calculate area.

**Let's learn**

A **composite shape** is a shape made up of two or more simple shapes.

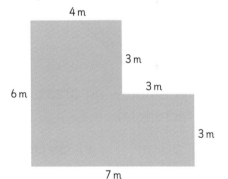

4 m

3 m

3 m

6 m

3 m

7 m

I'm having trouble calculating the area of this shape because it's composite. Can you help me?

4 m

3 m

3 m

7 m

We can split it up into regular parts like this and add their areas together. Can you find another way of splitting this shape?

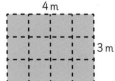

4 m

3 m

Finlay has made it easier to calculate the area of this composite shape by splitting it into two rectangles.

If we add the area of both together we will find the area of the whole shape.

Area = 3 m × 4 m

= 12 m²

Area = 3 m × 7 m

= 21 m²

The total area of the shape is:

**12 m² + 21 m² = 33 m²**

3 m

7 m

## Let's practise

1) Calculate the area of each section of the children's school playground design:

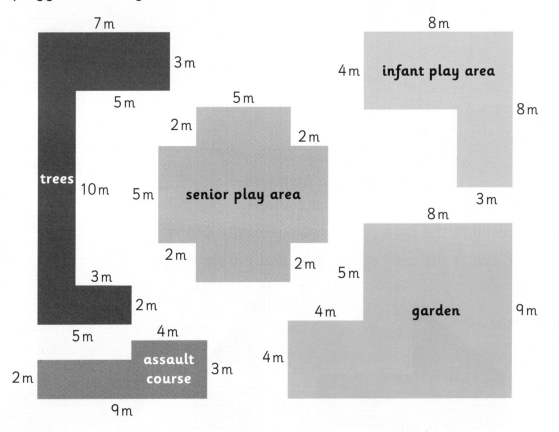

2) Use a ruler to measure the composite shapes below and calculate their surface area:

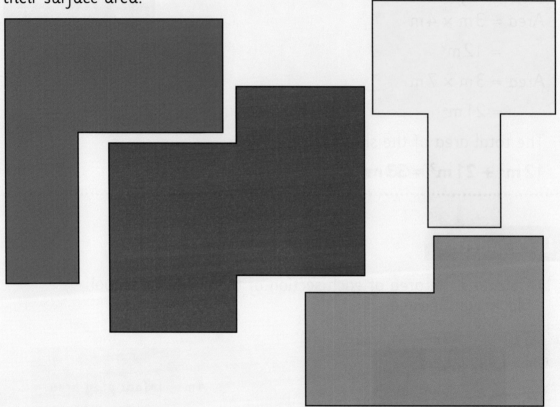

★ CHALLENGE!

Can it be split into two regular shapes?

i) Explain how Isla's idea can help find the area of the red composite shape.

ii) Calculate the area of the red composite shape.

iii) Calculate the area of the blue and purple shapes.

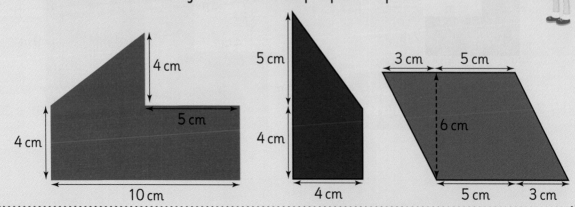

## 9.10 Calculating volume

We are learning to calculate the volume of simple cubes and cuboids.

**Before we start**

Here are the bases of two cuboids:
Each cube measures 1 cm³.
What would the volume of each cuboid be if there were six layers?

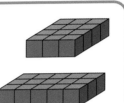

We can calculate the volume of a cube or cuboid without having to count cubes.

**Let's learn**

We can find the volume of a 3D shape by counting layers:

This cuboid is made up of three layers of 6 cm³:

**6 cm³ + 6 cm³ + 6 cm³ = 18 cm³**

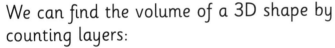

layer 3 →
layer 2 →
layer 1 →

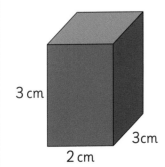

3 cm

3cm

2 cm

This cuboid is exactly the same size but hasn't been built using cubic centimetres. We can find the volume using the formula:

Volume = length × breadth × height

= 2 cm × 2 cm × 3 cm

= 18 cm³

## 9

1) Calculate the volume of the following shapes:

a)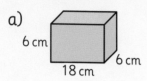
6 cm
6 cm
18 cm

b)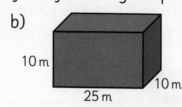
10 m
25 m
10 m

c)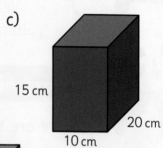
15 cm
20 cm
10 cm

d)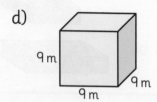
9 m
9 m
9 m

e)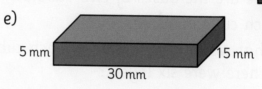
5 mm
30 mm
15 mm

2) The children are packing books into boxes for a classroom move:

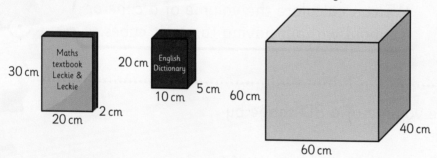

Maths textbook Leckie & Leckie
30 cm
20 cm
2 cm

English Dictionary
20 cm
10 cm
5 cm

60 cm
60 cm
40 cm

a) What is the volume of the:
   i) dictionary      ii) maths textbook      iii) box
b) How many maths textbooks can be packed into a box?
c) How many dictionaries can be packed into each box?

## CHALLENGE!

Each of these dominoes has dimensions double that of the previous one:

i)  Calculate the volume of each domino shown.
ii) Calculate the volume of the next two dominoes in this sequence.

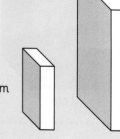

5 cm
2 cm
1 cm

## 9.11 Volume: composite shapes

We are learning to find the volume of composite cuboids.

**Before we start**

Calculate the volume of both these cuboids:

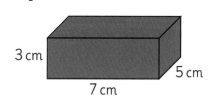

4 cm  8 cm  2 cm

3 cm  5 cm  7 cm

We can use regular shapes to calculate the volume of composite shapes.

**Let's learn**

A composite cuboid is a 3D shape made up of two or more cuboids:

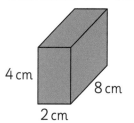

2 cm  4 cm  3 cm  2 cm  2 cm  2 cm

I'm not sure how to calculate the volume of this shape.

We can partition it into two cuboids and add their volumes together.

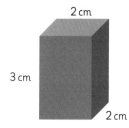

2 cm  3 cm  2 cm

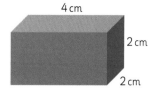

4 cm  2 cm  2 cm

Volume = 2 cm × 2 cm × 3 cm
$$= 12 \text{ cm}^3$$

Volume = 4 cm × 2 cm × 2 cm
$$= 16 \text{ cm}^3$$

Volume = length × breadth × height

Total volume = $12 \text{ cm}^3 + 16 \text{ cm}^3$
$$= 28 \text{ cm}^3$$

## Let's practise

1) Calculate the volume of the following shapes:

a)
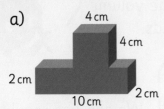

b)
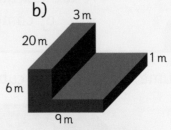

c)
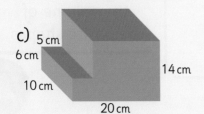

d)
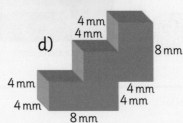

2) Which of these cuboids can be combined to make a composite shape with a total volume of:

a) 204 cm³   b) 135 cm³   c) 222 cm³   d) 96 cm³   e) 285 cm³

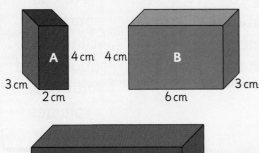

## ⭐ CHALLENGE!

i) Calculate the volume of each of these composite shapes:

ii) Draw a composite shape with a total volume of:

a) 100 cm³      b) 150 cm³

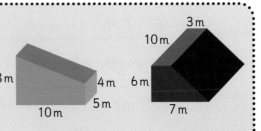

# 9 Measurement

## 9.12 Capacity problems

We are learning to solve problems involving capacity.

**Before we start**

What is the volume of each of these cuboids?

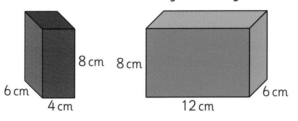

8 cm

8 cm

6 cm

6 cm

4 cm

12 cm

We can calculate capacity using length × breadth × height.

**Let's learn**

We can find the capacity of a container by working out the volume of cubic centimetres or metres needed to fill it and converting it to millilitres or litres:

This container needs nine cubic centimetres to fill it:

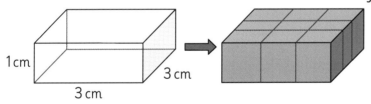

1 cm

3 cm

3 cm

Therefore, the capacity is 9 ml.

We can calculate this using the formula:

Capacity = length × breadth × height

$\qquad$ = 3 cm × 3 cm × 1 cm

$\qquad$ = 9 cm³ or 9 ml

**9**

We can calculate the capacity of this container using the same formula:

Capacity = length × breadth × height

        = 30 cm × 20 cm × 20 cm

        = 12 000 ml

We know that 1000 ml = 1 litre

So, the **capacity** of this container is **12 litres**.

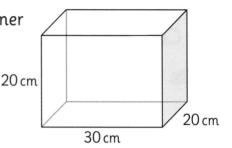

## Let's practise

1) Calculate the capacity of each of the following containers:

a)
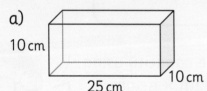
10 cm
25 cm
10 cm

b)
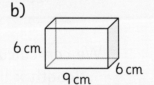
6 cm
9 cm
6 cm

c)
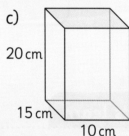
20 cm
15 cm
10 cm

d)
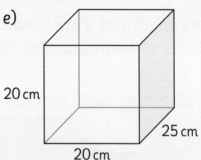
15 cm
20 cm
15 cm

e)
20 cm
25 cm
20 cm

2) The children have partly filled each of the containers shown with water. How much water will they have to pour into each so that they are completely full?

a)
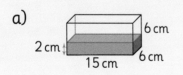
6 cm
2 cm
15 cm
6 cm

b)
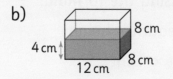
8 cm
4 cm
12 cm
8 cm

c)
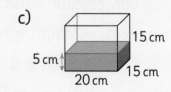
15 cm
5 cm
20 cm
15 cm

d)
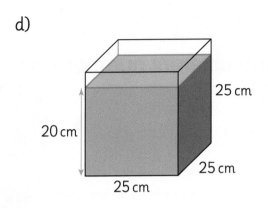
25 cm
20 cm
25 cm
25 cm

e)
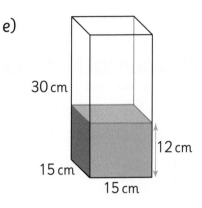
30 cm
12 cm
15 cm
15 cm

**CHALLENGE!**

Nuria is shopping for a fish tank. She has been advised that she needs a tank that holds more than 10 litres of water. Which of the following are suitable?

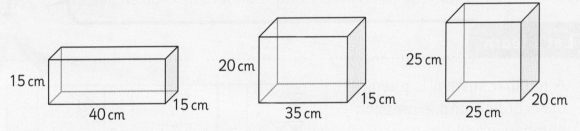

15 cm
40 cm
15 cm

20 cm
35 cm
15 cm

25 cm
25 cm
20 cm

The tank must contain 10% gravel and she intends to put in three decorations that have the following volumes:

200 cm³        350 cm³        450 cm³

i) Which tank should Nuria choose to fit in the water, gravel and decorations?

ii) How many different containers can you draw with a volume (capacity) of 24 cm³ (24 ml)?

# 10 Mathematics, its impact on the world, past, present and future

## 10.1 Mathematical inventions and different number systems

We are learning to investigate historical number systems and mathematical discoveries.

**Before we start**

List how many different ways you use mathematics within a day. Discuss with a partner the things you would NOT be able to do without using mathematics, and make a list to share.

Numbers and number systems are mathematical notations for representing numbers of a given set.

**Let's learn**

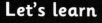

A number system is a way of recording and expressing numbers using digits, letters or symbols in a consistent way.

The Babylonian system appeared around 200 BC and is credited as being the first known positional numeral system, in which the value of a particular digit depends both on the digit itself and its position within the number.

The Babylonians did not have a digit for, nor a concept of, the number zero. They understood the idea and instead left a space as a placeholder.

| | | | | | |
|---|---|---|---|---|---|
| | 1 | | 11 | | 30 |
| | 2 | | 12 | | |
| | 3 | | 13 | | 40 |
| | 4 | | 14 | | |
| | 5 | | 15 | | 50 |
| | 6 | | 16 | | |
| | 7 | | 17 | | |
| | 8 | | 18 | | |
| | 9 | | 19 | | |
| | 10 | | 20 | | |

**Let's practise**

1) Using the information from the tables on the previous page, write down the following numbers using Babylonian numerals:

　　a) 5　　　　b) 22　　　c) 40
　　d) 59　　　e) 17　　　f) 34
　　g) 43　　　h) 51

2) Using the information from the tables on the previous page, write down the following facts about yourself and ask a friend to check these:

　　a) age　　　　　　　　　　　　b) shoe size
　　c) number of people in your family　　d) number of pupils in your class

3) Copy and complete the following calculations using Babylonian numerals:

　　a)  +  =

　　b)  +  =

　　c)  − =

**CHALLENGE!**

Let's investigate …

David Brewster is a famous Scottish mathematician.

Work with a partner to research what he is famous for and what impact he has had on the world.

Create a Fact File with the information you find to include the following details:

- His name and date of birth (you could include his nickname!).
- Where in Scotland he was from.
- What he was best known for.
- Examples of his work (include pictures and diagrams).

## 11.1 Applying knowledge of multiples, square numbers and triangular numbers to generate number patterns

> We are learning to apply knowledge of multiples, square numbers and triangular numbers.

**Before we start**

Each snowman needs the same amount of buttons, complete the following calculations using the information in the table.

| No. of snowmen (s) | 1 | 2 | 3 | 4 | 5 |
|---|---|---|---|---|---|
| No. of buttons (b) | 5 | 10 | 15 | ? | ? |

a) For each additional snowman, how many extra buttons are required?

b) Copy and complete the formula:

**Number of buttons = _____ × number of snowmen**

c) Copy and complete the formula using symbols:

**b = _____ × s**

d) Use the formula to work out how many buttons you would need for nine snowmen.

> A **number pattern** can contain **multiples**, **square numbers** and **triangular numbers**.

**Let's learn**

A **multiple** is a number that can be divided by another number without leaving a remainder.

   *e.g.*   **3, 6, 9, 12** … *these are all multiples of 3*

A **square number** is the product of a number multiplied by itself.

   *e.g.*   **1, 4, 9, 16** … *1 × 1 = 1, 2 × 2 = 4, 3 × 3 = 9, 4 × 4 = 16* …

A triangular number is any number that can make a triangular dot pattern.

   *e.g.* 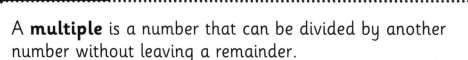    *1, 3, 6, 10… are all triangular numbers.*

**Let's practise**

1) Continue the following number sequences and write a rule for each one explaining the number pattern used and the multiple within the sequence:

   a) 12, 16, 20, 24, ___, ___, ___. Rule: _____ Multiple: __
   b) 27, 36, 45, 54, ___, ___, ___. Rule: _____ Multiple: __
   c) 32, 40, 48, 56, ___, ___, ___. Rule: _____ Multiple: __

2) Copy and complete the following table:

| No. (n) | 1 | 2 | 3 | 4 | 5 | 6 | 7 |
|---|---|---|---|---|---|---|---|
| Square number (s) | 1 | 4 | 9 | 16 | ? | ? | ? |

   a) Copy and complete the formula:
      square number = _____ × number
   b) Copy and complete the formula using symbols:
      s = _____ × n
   c) Use the formula to work out the 9th and 10th square numbers.

3) Copy and complete the following table:

| No. (n) | 1 | 2 | 3 | 4 | 5 | 6 | 7 |
|---|---|---|---|---|---|---|---|
| Triangular number (t) | 1 | 3 | 6 | 10 | ? | ? | ? |

   Can you work out the 8th and 9th triangular numbers?

**CHALLENGE!**

Finlay is making a Christmas tree using pom poms and he needs to work out how many he will need.

*He starts by drawing a diagram to help him ... he got to four rows and decided to find a different way to work it out!*

Using the table and the formula, can you help Finlay work out how many pom poms he will need?

| No. (n) | 1 | 2 | 3 | 4 | 5 | 6 | 7 |
|---|---|---|---|---|---|---|---|
| Triangular number (t) | 1 | 3 | 6 | 10 | ? | ? | ? |

   a) Work out how many pom poms Finlay will need if there are 10 rows.
   b) Work out how many pom poms Finlay will need if there are 15 rows.

# 12 Expressions and equations

## 12.1 Solving equations with inequalities

> We are learning to solve inequalities by balancing two number sentences.

**Before we start**

1) Using the information in the word problem, copy and complete the number sentence:

   a) Nuria's pet cats have 20 legs between them. How many cats does she have?

   $4 \times \boxed{\phantom{0}} = 20$

   b) Finlay went fishing and caught eight fish each day. For how many days would he have to fish to catch 32 fish?

   $8 \times \boxed{\phantom{0}} = 32$

2) Read the following word problems, write the number sentence and solve it:

   a) Amman has 28 sweets and shares them equally between four friends. How many sweets do they get each?

   b) Mrs Smith has 11 textbooks and 33 pupils. How many pupils will need to share each textbook?

> Equations are two number sentences that are connected by an equals sign showing each side is balanced.

**Let's learn**

An **equation** states that two things are equal and can contain numbers and pictures to represent the values.

Look at the scales; each side contains different shapes.

Each shape represents a different value.

To solve balance problems, you need to make the values on each side balance.

Therefore, **two squares, a circle and a triangle has the same value as four squares and a triangle.**

This means that **the circle has the same value as two squares**.

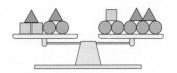

**Let's practise**

1) Write a sentence to describe each of the balance images. The one on the right has been done for you.

> *Two triangles, two squares and two circles is the same as two triangles, one square and five circles.*
>
> *This means that the square has the same value as three circles.*

a)

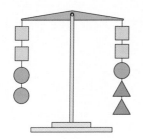

b)

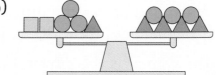

c)

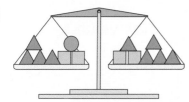

2) For these questions find the missing numbers:

a) Find ▢ if ▲ = 3

   ▢ =  ☐

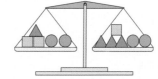

b) Find ▲ if ● = 4

   ▲ =  ☐

c) Find ⌐ if ☺ = 6

   ☺ =  ☐

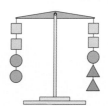

3) Work in pairs. Draw a balance in your maths jotter.

- Player A starts on any of the green circles on the left of the grid. Both players draw a circle on the left pan of their balance picture.

- Player B then picks a shape to the right, moving horizontally or diagonally. Both players draw this shape on the right pan of their balance picture.

- Continue until you get to the other side of the grid.

- When you get to the other side, you will have a balance problem. If the value of ⬤ is 30, what is the value of ▢?

- The winner is the player with the highest value of ▢.

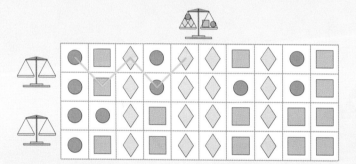

What is the value of ▢?

Explain the method you used to solve your balance problem.

**CHALLENGE!**

Find ▲ if ▢ = 2    ▲ = [ ]

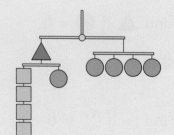

# 13 2D shapes and 3D objects

## 13.1 Describing and sorting triangles

We are learning to identify and sort different types of triangle.

**Before we start**

Finlay says this is a triangle. Do you agree? Give reasons for your answer.

There are three different types of triangle.

**Let's learn**

In a scalene triangle, all three sides are different lengths, and all three angles are different.

An isosceles triangle has two equal sides, two equal angles and a line of symmetry.

An equilateral triangle has three equal sides, three equal angles and three lines of symmetry.

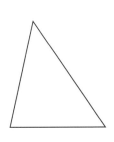

Scalene

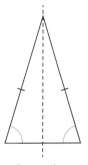

Isosceles

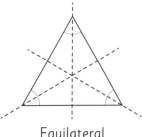

Equilateral

**Let's practise**

1) Sort these triangles by copying and completing the table. Use a protractor to check the angles. One has been done for you.

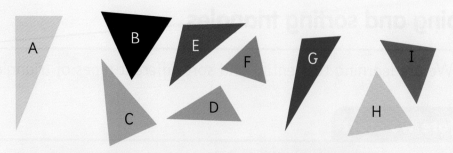

| Scalene | Isosceles | Equilateral |
|---------|-----------|-------------|
| A | | |

2) For each triangle, measure the missing angles to work out what kind of triangle it is.

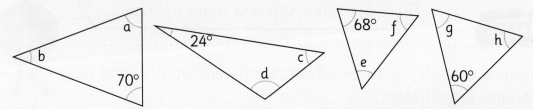

3) Use a ruler and protractor to draw the following triangles:
   a) A scalene triangle with one 45° angle.
   b) An isosceles triangle with a base that is 5 cm long.
   c) A scalene triangle with one side of 6 cm and one 130° angle.
   d) An equilateral triangle.

### CHALLENGE!

Investigate these triangles.

a) Can you find three different scalene triangles with one 5 cm side and one 8 cm side?

b) How many isosceles triangles can you find with one 5 cm side and one 8 cm side?

c) Can you find an equilateral triangle with one 5 cm side and one 8 cm side?

d) Are your answers different if two sides measure 5 cm?

## 13.2 Drawing 2D shapes

We are learning to draw 2D shapes according to their properties.

**Before we start**

Name the shapes that have parallel sides.

A    B    C    D    E

Use a ruler when drawing 2D shapes.

**Let's learn**

Congruent shapes are exactly the same shape and size, although they may appear different if one is rotated or positioned differently. These two rhombuses are congruent:

When we draw congruent shapes, we need to make sure that all the sides are exactly the same length, and all the angles are exactly the same.

**Let's practise**

1) Draw a picture that uses these shapes: triangles, rectangles, pentagons, hexagons and decagons. Make the rectangles congruent and the triangles congruent. Use a ruler to draw straight sides and a protractor to check the angles.

2) Name and sketch the shape that matches the description.

a) This shape has:
- eight straight sides of equal length
- eight vertices.

b) This shape has:
- six straight sides of different lengths
- six vertices.

c)  This shape has:
*   four straight sides
*   two pairs of parallel opposite sides.

d)  This shape has:
*   three straight sides of the same length
*   three vertices.

3)  Draw two examples of each shape. The first one has been done for you. They should not be congruent, but should be different in size and position.

| Rectangles | Rhombuses | Trapeziums | Parallelograms |
|---|---|---|---|
|  | | | |

**CHALLENGE!**

Use a protractor and a ruler to draw:

a)  Two congruent rhombuses ABCD with sides of 3 cm and where angle ABC is 137°.

b)  Two congruent parallelograms ABCD with sides of 2 cm and 5 cm, where angle BCD is 67°.

# 13 2D shapes and 3D objects

## 13.3 Making representations of 3D objects

We are learning to draw and make representations of prisms.

### Before we start

Amman says that a cylinder is a prism because it has two identical faces, and the cross-section is the same all the way through. Isla disagrees, she thinks a prism must have flat sides. Who is right? Explain your thinking.

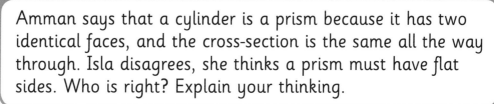

We can represent 3D objects by drawing or by making skeletal models.

### Let's learn

To draw a prism, draw the two identical faces first, one offset from the other. Join the faces with straight lines, using a ruler. Draw any hidden edges with a dotted line.

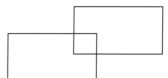

Triangular prism        Rectangular prism

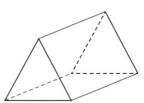

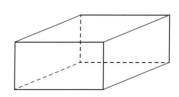

### Let's practise

1) Here are some faces of prisms. For each one, draw the prism.

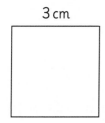

3 cm

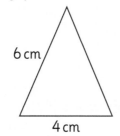

6 cm

4 cm

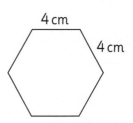

4 cm

4 cm

2) a) Use straws and modelling clay to construct the skeleton of each of these 3D objects.

A    B    C    D

b) Complete the table. Write the number of long and short straws and blobs of modelling clay you used to make each skeleton.

| Skeletal model | Number of | | |
|---|---|---|---|
| | long straws | short straws | blobs of sticky putty |
| A | | | |
| B | | | |
| C | | | |
| D | | | |

3) Look at the four skeletal models in Question 2. Write which models have:

a) three edges only at each vertex

b) more than three edges at one vertex

c) more than three rectangular faces.

**CHALLENGE!**

Leonhard Euler, a Swiss mathematician, said: 'The sum of the number of faces and vertices is always equal to the number of edges plus two'.

a) Is Euler's rule true or false? Investigate for these solids.

| 3D solid | Number of | | | |
|---|---|---|---|---|
| | faces (F) | vertices (V) | edges (E) | F + V |
| cube | | | | |
| hexagonal prism | | | | |
| square-based pyramid | | | | |
| pentagonal prism | | | | |
| triangular prism | | | | |
| pentagonal-based pyramid | | | | |

b) Does Euler's rule, F + V = E + 2, work for a cone? Give a reason for your answer.

# 13 2D shapes and 3D objects

## 13.4 Nets of prisms

We are learning to identify and make nets of prisms.

### Before we start

Finlay thinks this is a net of a cube.
Nuria doesn't think so. Who is right?
Explain your answer.

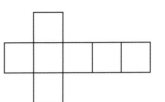

A net is a flattened 3D object. It can be made into the object by folding it.

### Let's learn

Here is the net of a regular triangular prism.
The end faces of this prism are equilateral triangles.

The net can be folded up to make a triangular prism:

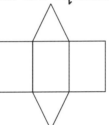 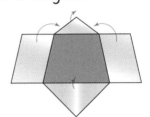 

### Let's practise

1) Name the 3D objects from their nets.

A

B

C

D

E

F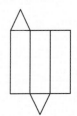

2) Three of these are nets of 3D objects.

   a) Name the object from the net.

   b) Say which diagram is **not** a net. Explain your thinking.

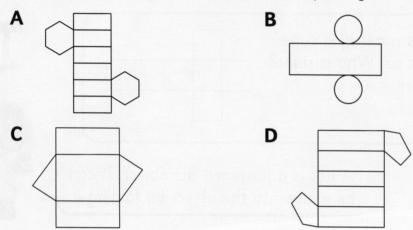

**A**                          **B**

**C**                          **D**

3) Sketch a net of an octagonal prism.

   Cut out your net and fold it to check.

## CHALLENGE!

A triangular prism has end faces that are isosceles right-angled triangles. The equal sides measure 4 cm.

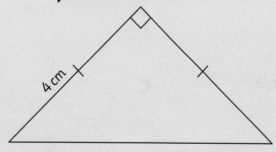

4 cm

The length of the prism is 5 cm.

a) Draw a net of the triangular prism.

b) How many different nets can you draw?

## 14.1 Drawing triangles

We are learning to draw triangles.

**Before we start**

Isla says that the base of an isosceles triangle is always shorter than its equal sides. Is she right? Explain your thinking and draw a diagram to prove your answer.

When we draw triangles, we measure the angles that are **inside** the shape – the **interior** angles.

**Let's learn**

**How to draw a triangle, given the sides and angles.**

Draw a line AB the length of the base.

Measure and mark the angle at A using a protractor. Draw line AC, the length of the first side.

Draw a line with a ruler from B to C.

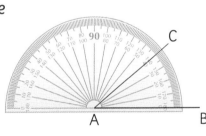

**Let's practise**

1) Draw these triangles accurately in your jotter. Measure the missing angles. (The diagrams are not to scale.)

a)

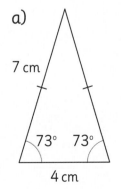

b)

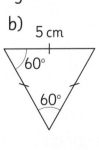

c)

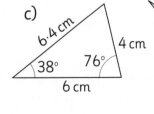

d)

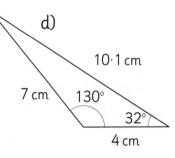

2) Draw the following triangles ABC, then measure the missing lengths and angles:

a)  AB = 5 cm, AC = 4 cm, BC = 3 cm.

Angle A is 38°

Angle B is 52°

Angle C is

b)  AB = 8.5 cm, AC is 4 cm, BC is 10 cm

Angle A is 100°

Angle B is 23°

Angle C is

c)  AB = 10 cm, AC = 6 cm.
Angle A is 33°

BC =                          cm

Angle B is

Angle C is

3) Draw your own triangles. Measure the lengths and the angles. Challenge a friend to draw your triangles accurately.

## CHALLENGE!

Start by drawing a triangle with AB = AC = 5 cm, and angle B = angle C = 45°.

Draw another triangle. Keep length AC as 5 cm and angle A as 90° but add 1 cm to length AB.

What happens to angles B and C?

Keep adding 1 cm to AB.

Investigate what happens to angles B and C each time you add 1 cm to AB. By how much do they change?

What happens if you start with a different triangle?

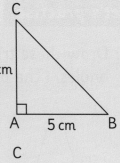

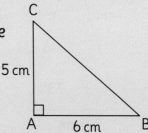

## 14.2 Reflex angles

We are learning to measure and draw reflex angles.

**Before we start**

What is the smallest reflex angle you can think of? What is the largest? Explain your answers.

When drawing reflex angles, make sure you mark the correct part of the angle.

**Let's learn**

Most protractors measure angles up to 180°.

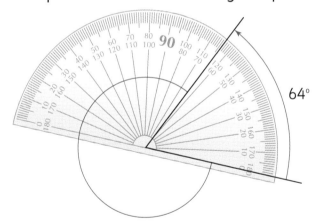

64°

To draw a reflex angle, calculate first, then draw the smaller angle.

For example, to draw a reflex angle of 198°, first calculate:
$360 - 198 = 162°$.

Now draw an angle of 162°.

Make sure you mark the reflex angle and not the angle you drew.

## Let's practise

1) Measure these reflex angles.

a)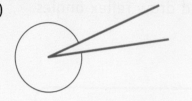

b)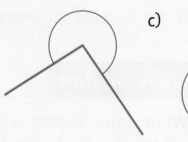

c)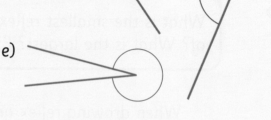

d)

e)

2) For each of these reflex angles, calculate the acute or obtuse angle that will help you draw the reflex angle.

a) 187°      b) 311°      c) 246°      d) 269°

3) Draw the following reflex angles in your jotter:

a) 270°      b) 315°      c) 199°      d) 278°      e) 221°

★ **CHALLENGE!**

Work with a partner. Each of you should write down 10 angles between 180° and 360°.

Challenge your partner to create a pathway design that uses these 10 reflex angles.

Measure your partner's angles to check. An example of a pathway diagram is shown in the diagram.

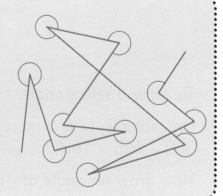

## 14.3 Finding missing angles

We are learning to find missing angles up to 360°.

**Before we start**

Amman says that if he measures two angles he knows the third. Nuria reckons she can measure one and calculate the other two. Who is right? Explain why.

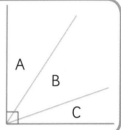

A
B
C

Angles around a point add up to 360°.

**Let's learn**

When we measure angles, we are measuring the length of the turn between the angle arms. These angles are all around the point A. Together they make a full turn, so they add up to 360°.

**Remember**

Complementary angles make a right angle. They add up to 90°.

Supplementary angles add up to 180°. Together the angles make a half turn.

70°
140°
A
150°

140° 40°

35°
55°

**Let's practise**

1) Calculate the missing angles.

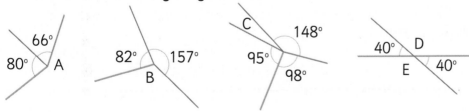

66°
80° A

82° 157°
B

C 148°
95°
98°

40° D
E 40°

2) Measure angle A. Calculate to find angle B.

a)

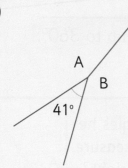

41°

b)

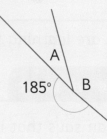

185°

c)

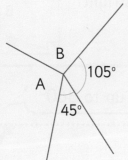

105°

45°

d)

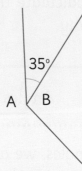

35°

3) Without measuring, calculate the missing angles.

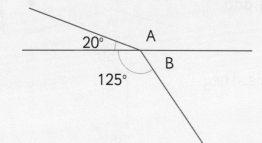

20°

125°

A

B

C

59°

E

D

---

⭐ **CHALLENGE!**

These isosceles triangles meet at a point. Calculate the missing angles.

Hint: Angles in a triangle add up to 180°.

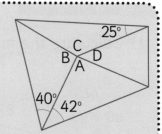

25°

C

B   D

A

40°   42°

---

# 14 Angles, symmetry and transformation

## 14.4 Using bearings 1

We are learning to use bearings to describe a journey.

**Before we start**

Finlay is standing facing north. He turns on a bearing of 215°. In which compass direction is he now facing?

We can use three-figure bearings to describe a unique route.

**Let's learn**

When we describe a journey using three-figure bearings, we take a new bearing each time we change direction.

For example, Nuria walks from her house to Isla's house. On the way she stops at Amman's.

She describes this journey by first taking the three-figure bearing to show the direction from her house to Amman's.

She writes

*Go 300 m on a bearing of 040°*

N

040°

300 m

111°

500 m

She then takes the bearing of Isla's house from where she is standing at Amman's.

She writes

*Go 500 m on a bearing of 111°*

*Remember – always measure bearings from north.*

## Let's practise

1) Finlay is tracking a bear through the woods. He takes this route:

   Describe Finlay's route using three-figure bearings.

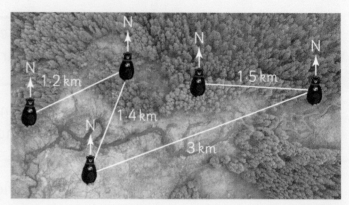

2) A plane is flying over the ocean when it changes course. This is the route it takes.

   Describe the plane's route using three-figure bearings.

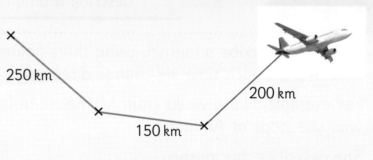

250 km  150 km  200 km

## CHALLENGE!

Nuria walks from A to B.

Amman walks from C to F via D and E.

Describe their journeys using three-figure bearings. Write distances in km using the scale given.

Make sure you measure from the **points** of the pins.

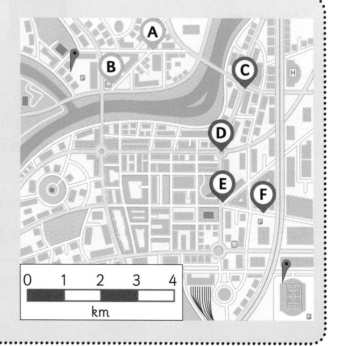

# 14 Angles, symmetry and transformation

## 14.5 Using bearings 2

We are learning to use bearings to plot a journey.

### Before we start

Nuria says that the bearing of B to C is 125°. Finlay says it is 75°. Who is right, and why?

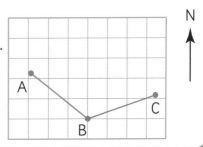

We can use three-figure bearings to plot a unique route.

### Let's learn

When we plot a journey with bearings, we need to measure both the **bearing** and the **distance**. The distance may need to be scaled down. For example, if the distance given is 1 km you might draw this as 1 cm, and state that the scale is 1 cm for every 1 km, or 1 cm: 1 km.

Here is a journey: Go 3 km on a bearing of 070°, then 2 km on a bearing of 155°. How to plot a journey:

1) Mark the starting point. Measure the bearing given using a protractor. Draw a line showing the distance given along this bearing.

2) Mark the end of the line you have drawn, and use this mark to measure the next bearing and distance. **Take care if the bearing is a reflex angle.**

3) Continue until the whole journey has been plotted.

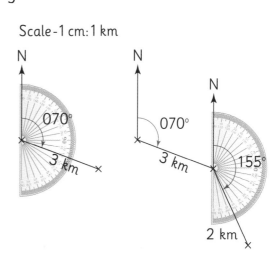

Scale – 1 cm : 1 km

1) Copy this grid.

   a) Starting at A, plot the following journey.
      Remember to start from north each time you reach a new point.

      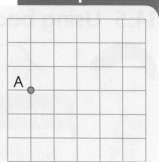

      Move two squares on a bearing of 045°.

      Move one square on a bearing of 090°.

      Move two squares on a bearing of 135°.

      Move three squares on a bearing of 225°.

      Move two squares on a bearing of 315°.

      Move one square on a bearing of 000°.

   b) Name the shape you have made.

2) A spider scuttles across a page. Plot its journey in your jotter:

   3 cm on a bearing of 217° *then* 4·5 cm on a bearing of 064° *then*.

   5 cm on a bearing of 158° *then* 7 cm on a bearing of 310°.

3) Isla is visiting a stone circle.
   She walks from stone to stone. Plot her route in your jotter, using a scale of 1 cm for 1 m.

   Each time you take a bearing, mark an x to show where the stones stand.

   2 m on a bearing of 065° *then* 1 m on a bearing of 090° *then*

   2 m on a bearing of 115° *then* 2 m on a bearing of 150° *then*

   4 m on a bearing of 240° *then* 3 m on a bearing of 285° *then*

   3 m on a bearing of 015°.

**CHALLENGE!**

Work with a partner.

Imagine and describe a journey using three-figure bearings. You can use distances in cm or you can choose your own scale.

Challenge your partner to plot your journey.

## 14.6 Using coordinates

We are learning to solve problems using coordinates.

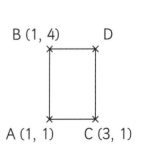

**Before we start**

Isla is plotting these coordinates, joining each one as she goes:
(1, 3) (4, 3) (4, 5) (1,5) (1, 3)
Predict what shape they will make. Explain your thinking.
Plot the coordinates for Isla to check.

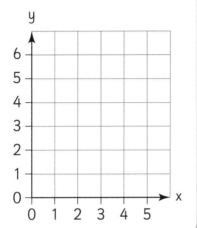

We can use our knowledge of the coordinate grid to help us solve problems.

**Let's learn**

Coordinates are given in the form (x, y), where the x coordinate tells you how far along the x-axis (horizontally) to go, and the y coordinate tells you how far along the y-axis (vertically) to go.

**Finding missing vertices**

Here is a rectangle. The coordinates of three of its vertices are given.

We can work out the missing vertex D by drawing the rectangle on the coordinate grid.

Alternatively, we can calculate D:

We know the shape is a rectangle, so the opposite sides are equal.

A and C are 2 units apart in the x direction, so B and D must also be 2 units apart.

Add 2 to the x coordinate of B (1, 4). This gives us an x coordinate of 3.

A and B are 3 units apart in the y direction, so C and D must also be 3 units apart.

Add 3 to the y coordinate of C (3, 1). This gives us a y coordinate of 4.

The coordinates of D are (3, 4)

## Let's practise

1) Plot these points on a coordinate grid to make part of the shape given. Write down the coordinates of the missing point or points.
   a) Rectangle: (1, 1) (1, 6) (4, 1)
   b) Square: (0, 1) (4, 0) (5, 4)
   c) Rhombus: (0, 3) (2, 0) (4, 3)
   d) Kite: (3, 1) (5, 5) (3, 6)

2) Calculate the missing coordinates for each shape.

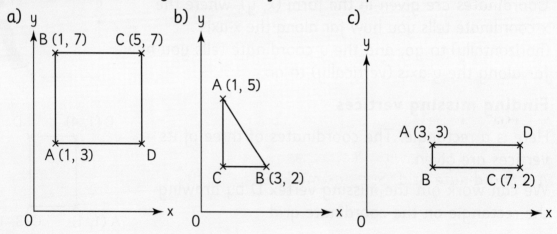

3) Each pair of shapes is congruent. Find the missing coordinates.

a)

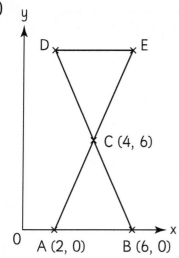

b)

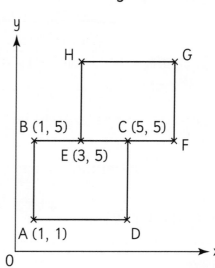

c)

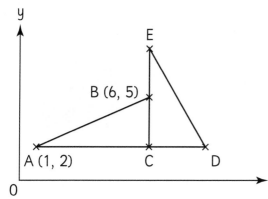

**CHALLENGE!**

Nuria has been drawing rectangles on a coordinate grid. She has made this pattern.

Nuria wonders what the coordinates of the vertices of the 10th rectangle will be.

Can you work it out without drawing?

Can you write a rule to find the coordinates of the vertices of any rectangle in the pattern?

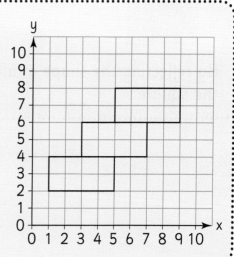

14.6 **Using coordinates**    197

## 14.7 Symmetry 1

We are learning to identify more than two lines of symmetry in shapes.

### Before we start

Isla folds a paper shape exactly in half, twice.
This is the shape she ended up with.
Draw her original shape.

Some shapes have more than two lines of symmetry.

### Let's learn

A square can be folded in half four different ways. It has two diagonal lines of symmetry and two vertical/horizontal lines of symmetry.

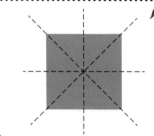

If you fold a square along all the lines of symmetry, you get this shape:

The folding stages are below.

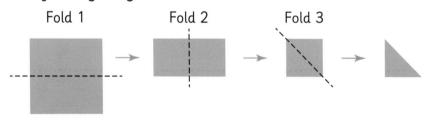

Fold 1        Fold 2        Fold 3

By reflecting the shape in all the lines of symmetry, you can recreate the square.

**Let's practise**

1) Copy each shape. Draw in all the lines of symmetry.

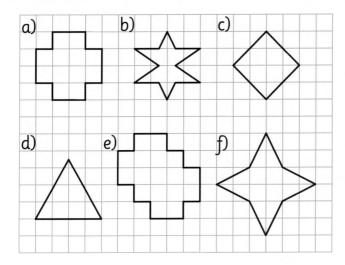

2) Sort these shapes according to their lines of symmetry. Put the letter of each shape in the correct place in the Venn diagram.

A)        B)        C)        D)

E)        F)        G)

H)        I)       J)

One or more diagonal line of symmetry

Vertical *and* horizontal lines of symmetry

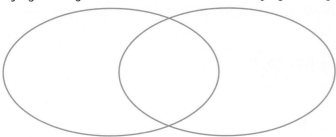

**CHALLENGE!**

Look at the following shapes.

A) 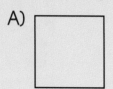    B)      C)     D)

E) 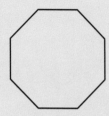    F)    G)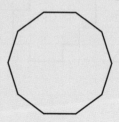

i) Copy and complete the table. One has been done for you.

|   | Name of shape | Number of lines of symmetry |
|---|---|---|
| A | Square | 4 |
| B |  |  |
| C |  |  |
| D |  |  |
| E |  |  |

ii) Use the table to predict how many lines of symmetry shapes F and G have.

iii) How many lines of symmetry are there in a regular polygon with 100 sides?

iv) What is a 100-sided polygon called?

## 14.8 Symmetry 2

> We are learning to complete symmetrical pictures with more than two lines of symmetry.

### Before we start

Amman reflects this shape in the first mirror line to make a new shape. He then reflects the new shape in the second mirror line to make a third shape. Draw each of Amman's shapes.

1 - - - - - - - - - - - - - - - - - - -

2 - - - - - - - - - - - - - - - - - - -

> To complete a picture or shape with four lines of symmetry, we have to reflect it diagonally, vertically and horizontally.

### Let's learn

This triangle has been reflected diagonally, then horizontally, then in the vertical line of symmetry.

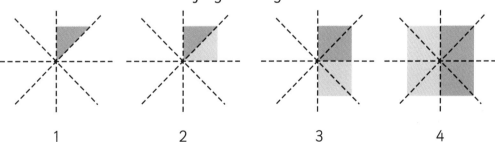

1              2              3              4

Does it matter which order you complete the reflections?

## Let's practise

1) a) Copy and complete the shapes by reflecting them in the lines of symmetry. Make sure the colours are symmetrical, too.

   b) Draw four lines of symmetry as shown and make your own design for a partner to complete.

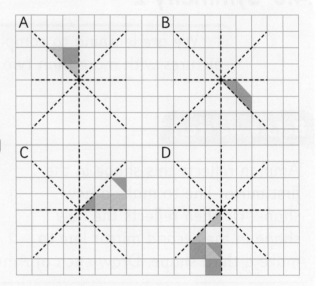

2) Copy and complete this design to give it four lines of symmetry.

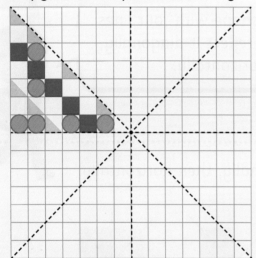

## CHALLENGE!

Create your own design in one of the triangles formed by the lines of symmetry.

Reflect it in all lines of symmetry to create a symmetrical square design, then reflect it again in the second vertical line of symmetry.

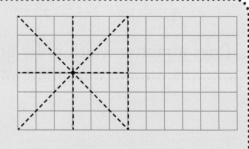

## 14.9 Using scale

We are learning to use scale to draw information on maps or plans.

**Before we start**

Isla knows that it is 30 km from Dundee to Kirriemuir. This is shown on her map as 1·5 cm.

On the same map, the distance from Dundee to Perth is 2 cm. How far is it in real life from Dundee to Perth?

To draw something onto a map or plan we need to know

1) the scale

2) the original distance.

**Let's learn**

**Take care**: the scale may be given in different units to the measurement taken.

Example:

Amman uses his family car's odometer (distance recorder) to find the distance between Cumbernauld and Paisley. It is approximately 30 km.

He wants to mark this distance on a map that has a scale of 1 cm : 5 km.

Because the scale is given in metres, Amman first has to convert 30 km into metres.

He converts 30 km to 30 000 m and then calculates:

$30\,000 \div 5000 = 6$, so the distance is 6 cm on his map.

Cumbernauld ✕

30 km

Paisley ✕

# 14

**Let's practise**

1) Finlay is driving with his parents around Italy. He has recorded the distances between some of the towns and cities.

| | Cities | Distance in km |
|---|---|---|
| a | Rome – Naples | 200 |
| b | Genoa – Milan | 120 |
| c | Turin – Milan | 130 |
| d | Florence – Pisa | 70 |
| e | Parma – Modena | 50 |

Finlay wants to draw a map to show where he has been.

He chooses a scale of 1 cm: 20 000 m.

Calculate, then draw a straight line to show each distance as it will appear on the map. Don't forget to label the cities.

One has been done for you:

Rome ✕————————————————————✕ Naples

2) Nuria is making a map of her local area. She has been out with a trundle wheel and measured some distances.

   i)   The **traffic lights** are 225 m west of the school gate.

   ii)  The **newsagent** is 130 m north of the traffic lights.

   iii) The **playpark entrance** is 120 m north-west of the school gate.

   iv) **Nuria's front door** is 70 m east of the newsagent.

   v)  **Amman's front door** is 85 m south-west of the playpark entrance.

   a)  Using a scale of 1 cm: 25 m, use a calculator to calculate the scaled distances.

   b)  Draw a rectangle 10 cm long and 6 cm high.

       Mark the school gate with an x at the bottom right-hand corner.

       Mark the different places on the map, using your scaled distances.

   c)  Measure the distance on your map from Nuria's front door to the playpark entrance. How far is it in real life?

3) A scientist is mapping the main craters of the moon. He knows that the moon is approximately 3500 km in diameter. He uses the scale 1 cm: 350 km.

a) Draw a circle to represent the moon using the scale given. Use a ruler to draw a line from the South to the North Pole.

b) Use a calculator to calculate the distances on the map using the scale given. Each one is measured from the South Pole.

Tycho crater is 350 km north then 175 km west.

Copernicus crater is 1750 km north then 875 km west.

Plato crater is 2975 km north then 560 km west.

c) Plot the location of the craters by marking them with an x.

d) The Sea of Tranquility is a feature of the moon. On the scientist's map it measures 1·8 cm across. How wide is the Sea of Tranquility really?

## CHALLENGE!

Make a map of your school playground.

Use a trundle wheel to measure distances accurately.

Decide on a suitable scale.

Plot the features accurately onto your map using the scale.

## 15.1 Sampling

We are learning to understand sampling in data collection.

### Before we start

Amman says this graph shows that children don't want to be on the library committee because they don't like reading. Is this a reasonable conclusion? Talk to a partner.

**Pupil committees**

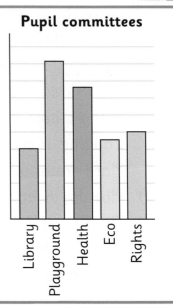

Sometimes, when we conduct a survey, we only gather data from some of the people who can take part. This is called a **sample** because it does not include everyone.

### Let's learn

Using a sample for a survey can give us a rough idea of what the answers from a whole population might be, it takes less time and is easier.

Consider the **big question** 'I wonder how children in my class travel to school?'

The whole population for this survey would be all the children in the class. A sample from the class might be 10 children.

Remember, a bigger sample will give a more accurate picture of the whole population than a small sample.

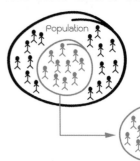

1) Decide whether each of these surveys gathers data from a whole population or from a sample. Explain your thinking.
One has been done for you.

a) On the 1st of August, members of the tennis club playing tennis are asked what could be done to improve the club. *This is a sample because it misses out people playing at the club on other days.*

b) All the staff at a restaurant are asked what food they like the best.

c) This week, pupils at my school were asked 'What do pupils at this school think of the playground?'

d) A TV company asks people watching a talent show programme to vote for their favourite performer by calling a special telephone number.

2) Copy and complete the table.

| | Big question | Whole population | Sample |
|---|---|---|---|
| a) | Where do the pupils in our school like to go on holiday? | All the pupils in the school. | 10 pupils from every class. |
| b) | What newspapers do the parents of pupils in our school read? | | |
| c) | How do the people who work in the local shops travel to work? | | |
| d) | What do librarians in Scotland read in their spare time? | | |

**CHALLENGE!**

Write two explanations for why the survey below DOES NOT gather data from all the people involved (population).

| Inquiry | Explanation |
|---|---|
| On Thursday, the children visiting a shop in a museum were asked: Which exhibit do children attending the museum like the best? | 1 |
| | 2 |

## 15.2 Interpreting graphs

We are learning to understand how changes in scale can influence the way a graph is interpreted.

**Before we start**

Which scale is most suitable for this data? Explain your answer.

| | Test score |
|---|---|
| Finlay | 30 |
| Nuria | 31 |
| Amman | 28 |
| Isla | 29 |

a)   80
     60
     40
     20

b)   40
     30
     20
     10

c)   35
     30
     25
     20
     15
     10
     5

The scale of a graph can be used to give a misleading message to the people reading the graph.

**Let's learn**

If you think a graph might be misleading, consider the **author**, the **audience** and the **purpose** of the graph. How might the author want their audience to interpret the graph?

Starting the scale at a point that is not zero makes the differences between groups look bigger.

### Points gained on sports day

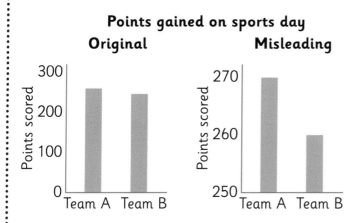

Making the scale on the y-axis longer than it needs to be makes the differences between points look less. In this example it makes the growth look smaller.

### Attendance at Badminton club

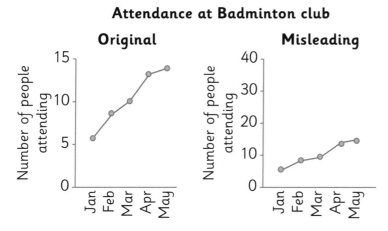

1) For each of these graphs, write whether it is accurate or misleading and say why.

a)

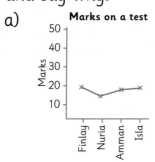

b)

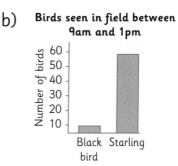

c) **People attending the gym on Tuesday**

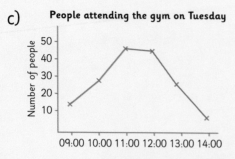

d) **Matches won this year**

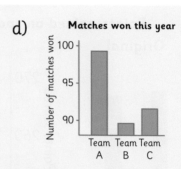

2) The Sunnytown Herald newspaper sold 443 612 copies this week. Its rival paper, the Sunnytown Express, sold 412 590 copies. These two graphs both represent this data.

(i)

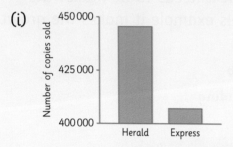

(ii)

a) Which graph is misleading? Explain how it is misleading.

b) Which of the papers is most likely to have presented the data in a misleading graph? Suggest a reason why they might have done this (think about **audience** and **purpose**).

3) A salesman shows his customer this graph of sales from a rival company over the past five months.

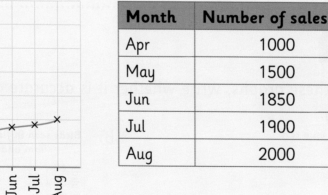

| Month | Number of sales |
|-------|-----------------|
| Apr | 1000 |
| May | 1500 |
| Jun | 1850 |
| Jul | 1900 |
| Aug | 2000 |

a) In your jotter, use the table to draw a graph using a scale that goes from 0 to 2000 using appropriate intervals.

b) Compare your graph with the salesman's graph. What do you notice?

c) What was the purpose of the misleading graph? How do you think the salesman wanted his audience to interpret his graph?

**CHALLENGE!**

A scientist is paid by a company to research a new fertilizer she has invented, Super-Gro.

She tests the fertilizer by measuring plant growth using Super-Gro and comparing this to Growalot, the current best fertilizer on the market.

Here are her results:

|  | Growth in mm | | |
|---|---|---|---|
|  | Day 1 | Day 2 | Day 3 |
| Super-Gro | 136 | 139 | 141 |
| Growalot | 135 | 136 | 138 |

a) Imagine you are the scientist. Create a (misleading) graph to convince people that Super-Gro is **much** better than Growalot, so they will buy it.

b) Give two reasons why the scientist is not the right person to be researching the fertilizer.

## 15.3   Understanding sampling bias

> We are learning about bias in sampling.

### Before we start

Nuria and Isla want to know where people in their street go on holiday.

Say what the whole population is for this survey. Suggest a sample they could use.

> Using a sample is easier and quicker than surveying a whole population, but we need to be careful about how we choose our sample.

### Let's learn

If we want to know 'What are the favourite sports of pupils in my school?' we will not get a good idea of the whole population if our sample only includes pupils in the football team. This is because the football team are more likely than other pupils to choose football as their favourite sport. This is called **bias**.

We can reduce bias by making our sample more representative of the whole population. Selecting the people to be sampled at random, for example by asking every third person on the register, is a good way to do this.

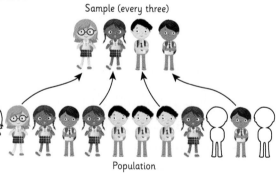

Sample (every three)

Population

### Let's practise

1) For each of these big questions, say whether the sample is biased or unbiased.

    a)   What sport do people at my school like doing?

        Sample: every fourth name from the register for each class.

b) Should a school spend £1000 on new computers or on musical instruments?
Sample: people in the school band.

2) Discuss with a partner whether or not these samples are good representations of the whole population. Explain your thinking in your jotter.

| | Big question | Sample |
|---|---|---|
| a) | What is the favourite ice-cream flavour of people who visit an ice-cream shop? | Every tenth person who visits the ice-cream shop in a week. |
| b) | How many mobile phones did a store sell in October? | All the mobile phones sold on Tuesdays in October. |
| c) | What toys do children in the nursery like to play with? | My little sister and her friends. |

3) Nuria conducts a survey to find out how many people in her school community watch the news on TV.

She doesn't think younger children will watch the news, so she decides to survey everyone in Primary 6 and 7.

They don't think younger children will watch the news, so they decide to survey everyone in Primary 6 and 7.

a) Give two reasons why this is not a good sample.
b) Suggest a more appropriate sample for them to use.

**CHALLENGE!**

Conduct your own inquiry. Decide on a big question, then consider how to collect your data.

*Will you collect a sample? If so, how will you avoid bias?*

Consider how to display your data, making sure your graphs are not misleading.

Carry out your enquiry.

## 15.4 Pie charts

We are learning to display and interpret data in pie charts.

**Before we start**

In one week, 70 people failed to turn up to their doctor's appointments. 300 appointments were made. What percentage of people attended their appointments? Give your answer to two decimal places.

Pie charts show data in relation to a whole population, and usually show percentage data.

**Let's learn**

In a pie chart, the whole population, or 100%, is represented by 360°.

1800 people are asked if they belong to a political party. 280 people belong to the Red party.

First calculate the angle $\frac{280}{1800} \times 360° = 56°$

Red party

16%

Then calculate the percentage $\frac{280}{1800} \times 100 = 16\%$ to the nearest whole percentage.

1) Finlay and Nuria keep a record for one week of the types of fruit sold in their healthy tuckshop.

| Fruit | Number of pieces sold | Number of degrees | Percentage of the total |
|---|---|---|---|
| Apple | 126 | | |
| Melon | 108 | | |
| Pear | 72 | | |
| Satsuma | 54 | | |
| Total | 360 | | |

a) Copy and complete the table.
b) Draw the pie chart. Label it with the percentages.
c) Write two things you notice about your pie chart.

2) A recruitment company wants to know how many people do the different jobs at Glasburgh airport. The airport provides them with this data:

| Job | Number of people | Degrees | Percentage |
|---|---|---|---|
| Ground staff | 864 | | |
| Passport inspector | 72 | | |
| Pilot | 216 | | |
| Flight attendant | 288 | | |
| Total | 1440 | 360 | 100 |

a) Copy and complete the table.
b) Draw the pie chart.
c) Write three statements about the data you have displayed.

**15**

A department store manager wants to know how successful his different departments are at selling their goods. He displays the data in a 3D pie chart.

**Sales by department**

a) List the departments in order from the one that sold the most to the one that sold the least.

b) The shoes department decide to redraw the pie chart.

The manager gives them the sales data.

| Department | Percentage sales | Degrees (to nearest whole degree) |
|---|---|---|
| Scarves | 53% | 191 |
| Shoes | 21% | 76 |
| Jewellery | 19% | 68 |
| Bags | 7% | 25 |

Use the table to redraw the pie chart. Label the sections with the percentages.

c) Write two statements comparing the 3D pie chart with the new pie chart. Why do you think the shoes department wanted to redraw the chart?

# 16 Ideas of chance and uncertainty

## 16.1 Predicting and explaining the outcomes of simple chance situations and experiments

We are learning to predict and explain the outcomes of simple chance situations and experiments using appropriate vocabulary and data.

### Before we start

Using the terms 0, 25%, 50%, 75% and 100%, work with a partner to investigate the following (*giving details of your choice of answer*):

a) What is the probability that June will follow July?

b) What is the probability that you will find £5 on the pavement?

c) What are the chances of all pupils in your class being boys?

d) What are the chances that there is sand on the beach?

The probability of something happening is the number of times it could happen within the total number of possible outcomes.

### Let's learn

When we talk about the **chance** or the **probability** of something happening, we discuss this on a likelihood scale between 0 and 1.

| 0 | 0·25 | 0·5 | 0·75 | 1 |
|---|------|-----|------|---|
| 0% | 25% | 50% | 75% | 100% |
| 0 | $\frac{1}{4}$ | $\frac{1}{2}$ | $\frac{3}{4}$ | 1 |

If we look at the chances of rolling a 1 on a dice …

There are six numbers on a dice, so we have $\frac{1}{6}$ of a chance to roll a number 1. This would be a less than 25% chance.

If we look at rolling an even number on a dice …

There are six numbers – 1, 2, 3, 4, 5 and 6, of which three are even. This is $\frac{1}{2}$ of the numbers, so we would have a 50% chance.

**Let's practise**

Using the appropriate vocabulary and numerical notation, answer the following questions:

1) What is the probability of Amman

   a) rolling a 4 on a dice?          b) rolling a 3 or a 4 on a dice?

2) If Amman has two dice:

   a) What is the likelihood that he will roll an odd number?
   b) If he rolls both dice three times, what are the chances that he will roll a 6?

3) If a bowl of fruit contains six apples, four pears and two oranges, what are the chances Isla will choose:

   a) a pear?          b) an orange?          c) a peach?

⭐ **CHALLENGE!**

Using the appropriate vocabulary and numerical notation, answer the following question:

Mrs Irwin has a box of 80 coloured pencils.

Some are red, some are blue, some are green and some are yellow.

There is a one in eight chance of choosing either a yellow or green pencil.

There is a one in four chance of choosing a blue pencil.

There is a one in two chance of choosing a pink pencil.

1)  How many yellow pencils are there?

2)  How many green pencils are there?

3)  How many blue pencils are there?

4)  How many pink pencils are there?

# Answers

## 1 Estimation and rounding

### 1.1 Rounding whole numbers (p.2)

**Before we start**

Biggest is 987 652
Digits represent 900 000, 80 000, 7000, 600, 50 and 2
Smallest is 256 789
Digits represent 200 000, 50 000, 6000, 700, 80 and 9

**Questions**

1) a) 478 939 lies between 470 000 and 480 000, rounded 480 000
   b) 847 235 lies between 840 000 and 850 000, rounded 850 000
   c) 765 812 lies between 760 000 and 770 000, rounded 770 000
   d) 38 036 lies between 30 000 and 40 000, rounded 40 000

2) a) 600 000    b) 800 000    c) 400 000
   d) 200 000    e) 800 000    f) 200 000

3) a) Incorrect. 390 000    b) Correct.
   c) Incorrect. 80 000      d) Incorrect. 400 000
   e) Correct.               f) Incorrect. 470 000
   g) Correct.               h) Incorrect. 710 000

**Challenge**

Answers will vary.

### 1.2 Rounding decimal fractions (p.4)

**Before we start**

Seven point four one
7·4 to the nearest $\frac{1}{10}$
7 to the nearest whole number

**Questions**

1) a) 5·87    b) 18·47    c) 7·81
   d) 0·01    e) 5·00     f) 0·94

2) a) false   b) false   c) true   d) false   e) false

3) a) 6·98 kg  b) 793·87 km  c) 0·02 litres  d) 8·01 kg

**Challenge**

2·35 km, 1·78 km and 0·93 km. Total 5·06 km
Total without rounding is 5·058 km. Estimate was very accurate.

### 1.3 Using rounding to estimate the answer (p.6)

**Before we start**

Isla is wrong it should be 390 000.
Isla has rounded to the nearest 1000.

**Questions**

1) a) Dana   Yes    b) Hamad   Yes    c) Jameela   No

2) a) **Not reasonable**. 15 000 + 45 000 = 60 000
   b) **Reasonable**. 90 000 − 5000 = 85 000
   c) **Not reasonable**. 45 × 20 = 9000
   d) **Not reasonable**. 4900 − 3900 = 1000
   e) **Reasonable**. 35 × 20 = 700
   f) **Not reasonable**. 400 + 300 + 1900 = 2600
   g) **Not reasonable**. 380 000 − 220 000 = 160 000
   h) **Reasonable**. 80 × 80 = 6400

3) a) Approximately £800 each.    b) Approximately £600 each.
   c) Approximately £240 each.

**Challenge**

Answers will vary.

## 2 Number – order and place value

### 2.1 Reading and writing whole numbers (p.8)

**Before we start**

a) Any three of:
   32 068    38 062    23 068    28 063    83 062    82 063
b) Any three of:
   26 308    26 038    23 608    23 068    20 638    20 368

**Questions**

1) a) one hundred and twenty two thousand, nine hundred and thirteen
   b) eight hundred and eight thousand, seven hundred and eight
   c) four hundred thousand and twenty-six
   d) one hundred and sixty-seven thousand and ninety
   e) three hundred thousand and one
   f) nine hundred and ninety-nine thousand, nine hundred and ninety-nine

2) a) 403 561    b) 399 240    c) 200 006
   d) 720 018    e) 972 911    f) 500 200

3) a) 693 250 sixty hundred and ninety-three thousand, two hundred and fifty
   b) 755 802 seven hundred and fifty-five thousand, eight hundred and two
   c) 542 346 five hundred and forty-two thousand, three hundred and forty-six

**Challenge**

8 236 477 eight million, two hundred and thirty-six thousand, four hundred and seventy-seven

### 2.2 Representing and describing whole numbers (p.10)

**Before we start**

a) 39 870    b) 37 089    c) 97 803 or 97 083

**Questions**

1) a) forty thousand, 40 000         b) four thousand, 4000
   c) six hundred thousand, 600 000  d) one, 1
   e) three hundred thousand, 300 000  f) ten thousand, 10 000
   g) thirty, 30                     h) eighty thousand, 80 000

2) a) 500 204 and number correctly represented by place value arrow cards or counters.
   b) 100 009 and number correctly represented by place value arrow cards or counters.
   c) 620 100 and number correctly represented by place value arrow cards or counters.
   d) 217 040 and number correctly represented by place value arrow cards or counters.

3) Answers will vary.

**Challenge**

There are four possibilities:
222 110    212 210    122 210    221 210

### 2.3 Place value partitioning of whole numbers (p.12)

**Before we start**

a) 60 hundreds **OR** 600 tens **OR** 6000 ones
b) 49 hundreds and 3 ones **OR** 490 tens and 3 ones **OR** 4903 ones
c) 27 hundreds and 6 tens **OR** 276 tens **OR** 2760 ones
d) 13 hundreds 5 tens and 8 ones **OR** 135 tens and 8 ones **OR** 1358 ones

## Questions

1) a) 798 thousands, 8 hundreds, 5 tens, 1 one
    798 thousands, 8 hundreds and 51 ones
    798 thousands, 85 tens and 1 one
    798 thousands and 851 ones

    b) 404 thousands, 9 hundreds, 6 tens, 9 ones
    404 thousands, 9 hundreds and 69 ones
    404 thousands, 96 tens and 9 ones
    404 thousands and 969 ones

    c) 260 thousands, 5 hundreds, 8 tens, 6 ones
    260 thousands, 5 hundreds and 86 ones
    260 thousands, 58 tens and 6 ones
    260 thousands and 586 ones

    d) 435 thousands, 2 hundreds, 1 ten, 8 ones
    435 thousands, 2 hundreds and 18 ones
    435 thousands, 21 tens and 8 ones
    435 thousands and 218 ones

2) a) 210348 partitioned in six different ways. For example
    $200000 + 10000 + 300 + 40 + 8$
    $210000 + 340 + 8$
    $210000 + 300 + 48$
    $210000 + 348$
    $210300 + 48$
    $210340 + 8$

    b) to f) Answers will vary but should follow the pattern used in a)

## Challenge

Amman has written 92000 but he should have written 920000. There are several possible answers to the second part of this question, including:

$900000 + 20000 + 1000 + 700 + 10 + 3$ or $921000 + 700 + 10 + 3$

## 2.4 Number sequences (p.14)

### Before we start

Top row:     14113, 14213, 14513, 14813
Middle row: 15013, 15413, 15613
Bottom row: 16013, 16113, 16213, 16313, 16613, 16713, 16913

### Questions

1) a) 218379   218380   218381   218382   218383
   b) 779999   780000   780001   780002   780003
   c) 800010   800011   800012   800013   800014
   d) 611598   611599   611600   611601   611602
   e) 305000   305001   305002   305003   305004
   f) 199999   200000   200001   200002   200003

2) a) 321501   321500   321499   321498   321497
   b) 578233   578232   578231   578230   578229
   c) 999996   999995   999994   999993   999992
   d) 610100   610099   610098   610097   610096
   e) 128001   128000   127999   127998   127997
   f) 400003   400002   400001   400000   399999

3) a) Increasing by 100000     b) Decreasing by 100
   c) Increasing by 10000      d) Decreasing by 10

## Challenge

a) 1000004   1000005   1000006   1000007   1000008
b) 8000000   9000000   10000000   11000000   12000000
c) 2000700   2000800   2000900   2001000   2001100

## 2.5 Comparing and ordering whole numbers (p.16)

### Before we start

a) There are 12 possible answers:

    17009 < 19700    17009 < 17900    17009 < 19070
    19700 > 17009    19700 > 17900    19700 > 19070
    17900 > 17009    17900 < 19700    17900 < 19070
    19070 > 17009    19070 < 19700    19070 > 17900

b) 17009    17900    19070    19700

## Questions

1) a) <        b) >        c) <        d) <
   e) <        f) >        g) >        h) <

2) a) False. Words should read eight hundred and four thousand, six hundred.
   b) False. Any number greater than 500090 written in words.
   c) False. Any number smaller than 113311 written in words.
   d) True        e) True

3) a) Any three pairs of numbers, all of which are multiples of ten. One number in each pair should be < 542135. The other should be > 542135.
   b) Answers will vary.

4) a) 233417   233419   233421   233424   233428
   b) 489656   489667   489682   489759   489761
   c) 941010   944009   944020   944023   949016

5)
| Pluto | 2390 km |
|---|---|
| Mercury | 4876 km |
| Mars | 6794 km |
| Venus | 12107 km |
| Earth | 12755 km |
| Neptune | 49527 km |
| Uranus | 51117 km |
| Saturn | 120536 km |
| Jupiter | 142983 km |

6) a) Answers should be six-digit numbers containing the digits 2, 1, 8, 0, 5 and 9, which meet the following criteria:
   - an even number – 0, 2 or 8 in the ones place
   - a multiple of 5 – 0 or 5 in the ones place
   - an odd number – 1, 5 or 9 in the ones place
   - a multiple of 10 – 0 in the ones place
   - largest possible number – 985210
   - multiple of 25 – number ending in 25 or 50

   b) Numbers made in Part a) written in order from largest to smallest.

## Challenge

a) 221645    221654    221556    112645    112654
   112556    212645    212654    212556

b) Answers will vary.

c) Ascending order: 112556, 112645, 112654, 212556, 212645, 212654, 221556, 221645, 221654
   Descending order: 221654, 221645, 221556, 212654, 212645, 212556, 112654, 112645, 112556

## 2.6 Negative numbers (p.20)

### Before we start

a) Fifty-two degrees Celsius and 52°C
b) Minus four degrees Celsius and –4°C

### Questions

1) Helsinki, Moscow, North Pole, Reykjavik
2) 22°C
3) 9°C
4) Moscow
5) Sydney
6) 16°C
7) 39°C

## Challenge

In ascending order, the depths are:
–40 m, –32 m, –24 m, –20·5 m, –18 m, –10 m, –9·2 m, –6m

## 2.7 Reading and writing decimal fractions (p.22)

**Before we start**

Answers will vary.

**Questions**

1) b) eighteen point four two one
   c) thirty seven point zero three four
   d) zero point six nine six
   e) twenty one point zero zero seven
   f) three hundred point zero zero four
   g) four hundred and ninety point two five five
   h) eight hundred and three point five zero six
   i) two hundred and sixty seven point four eight nine

2) a) 5·712     b) 0·136     c) 64·505
   d) 11·009     e) 95·080

3) a) Raikkonen took 3 whole seconds and 652 thousandths of a second longer than Vettel to complete the race.
   b) Bottas took 8 whole seconds and 883 thousandths of a second longer than Vettel to complete the race.
   c) Ricciardo took 9 whole seconds and 500 thousandths of a second longer than Vettel to complete the race.
   d) Hülkneberg took 28 whole seconds and 220 thousandths of a second longer than Vettel to complete the race.

**Challenge**

Nuria is incorrect because 9·950 means 9 whole seconds and 950 thousandths of a second or 9 whole seconds and 95 hundredths of a second.

## 2.8 Representing and describing decimal fractions (p.24)

**Before we start**

Each model represents the decimal fraction 2·03 because each shows two whole units (the units being a fully shaded rectangle made up of 100 parts, a £1 coin and a box of 100 pencils) and three hundredths of the same unit.

**Questions**

a) 1 one, 3 tenths, 5 hundredths and 2 thousandths
   $= 1·352 = 1\frac{352}{1000}$

b) 2 ones, 2 tenths, 0 hundredths and 6 thousandths
   $= 2·206 = 2\frac{206}{1000}$

c) 3 ones, 2 tenths, 8 hundredths and 3 thousandths
   $= 3·283 = 3\frac{283}{1000}$

d) 4 ones, 0 tenths, 0 hundredths and 4 thousandths
   $= 4·004 = 4\frac{4}{1000}$

e) 3 ones, 3 tenths, 1 hundredth and 0 thousandths
   $= 3·310 = 3\frac{310}{1000}$

**Challenge**

Answers will vary.

## 2.9 Zero as a placeholder in decimal fractions (p.26)

**Before we start**

Isla is incorrect. There are 10 possible answers. They are:

570 000     750 000     507 000     705 000     700 500
500 700     700 050     500 070     700 005     500 007

**Questions**

1) a) The placeholder is in the ones place.
   b) The placeholders are in the ones place and the tenths place.
   c) The placeholders are in the tenths place and the hundredths place.
   d) The placeholder is in the hundredths place.
   e) The placeholders are in the tens place and the tenths place.
   f) The placeholders are in the tenths place and the hundredths place.
   g) The placeholders are in the ones place and the hundredths place.
   h) The placeholders are in the tens, ones and thousandths places.

2) a) False → 2·4 = $2\frac{4}{10}$     b) True
   c) False → 12·03 = $12\frac{3}{100}$     d) True
   e) True     f) True

3) Answers will vary, but diagram/model should clearly show that:
   a) 0·4 = 0·40     b) 0·5 + 0·09 = 0·59
   c) 0·2 + 0·07 = 0·27     d) 0·8 + 0·01 = 0·81

**Challenge**

3·40 = 3·4          0·07 ≠ 0·70          0·29 = 0·290

## 2.10 Partitioning decimal fractions (p.28)

**Before we start**

All of the children are correct.

**Questions**

1) a) 5·126     b) 8·349     c) 1·766
   d) 4·682     e) 0·571     f) 0·053

2) a) 3 + 0·8     b) 5 + 0·5
   c) 7 + 0·9     d) 2 + 0·6
   e) 4 + 0·5 + 0·02     f) 8 + 0·3 + 0·01
   g) 1 + 0·1 + 0·07     h) 6 + 0·2 + 0·09
   i) 9 + 0·07     j) 2 + 0·2 + 0·04 + 0·006
   k) 7 + 0·4 + 0·003     l) 3 + 0·09 + 0·008
   m) 6 + 0·009

3) a) 352·3 and $352\frac{3}{10}$     b) 6·07 and $6\frac{7}{100}$
   c) 1·84 and $1\frac{84}{100}$     d) 4·009 and $4\frac{9}{1000}$
   e) 0·661 and $\frac{661}{1000}$     f) 7·105 and $7\frac{105}{1000}$

**Challenge**

Amman and Nuria are both correct.

## 2.11 Comparing and ordering decimal fractions (p.31)

**Before we start**

Nuria is incorrect. She has based her decision purely on the number of digits without considering their values.

The correct order is:

406     61·4     46     6·4     1·6     0·4

**Questions**

1) a) 1·049     1·094     4·091     4·109     9·104
   b) 3·629     3·926     6·926     9·326     9·926
   c) 1·255     2·515     5·125     5·255     5·552
   d) 0·689     0·869     6·986     6·989     8·009

2) a) The possibilities are: 2·41, 2·42, 2·43, 2·44, 2·45, 2·46, 2·47, 2·48, 2·49
   b) Answers will vary.

3) a) false     b) false     c) false     d) false
   e) false     f) false     g) true     h) true

4) a) 13·0 = 13     b) 1·89 > 1·8     c) 5·18 < 5·81
   d) 6·4 = 6·40     e) 3·09 < 3·9     f) 24 > 20·4

5) a) 26·5     26·05     25     2·65     2·5     2·06
   b) 178     17·8     7·81     7·08     1·78     1·7
   c) 43     34     9     4·09     3·49     0·94

6) a) 8 hundredths and 6 thousandths = 86
      thousandths = 0·086
   b) 7 hundredths and 2 thousandths = 72
      thousandths = 0·072

c) 3 tenths and 6 hundredths and 0 thousandths
   = 36 hundredths = 0·36

d) 500 thousandths = 50 hundredths = 5 tenths = 0·5

7) 0·57    0·765    1·6    1·751    6·175    6·51    7·5

## Challenge
196 tenths

# 3 Number – addition and subtraction

## 3.1 Mental addition and subtraction (p.34)

### Before we start
*NB: Other mental strategies are possible*

a) The best advice for Finlay is to use 'round and adjust' strategy.
   796 + 419 = 800 + 415 = 1215

b) The best advice for Finlay is to transform the calculation by adding the same amount to each side. 5062 − 37
   = 5065 − 40 = 5025

c) The best advice for Finlay is to count on from 988. 998 + 12 = 1000 + 7000 = 8000, so the answer is 8000 − **7012** = 988

### Questions
1) a) 17 357    b) 81 623    c) 77 628    d) 12 640
   e) 17 225    f) 31 998    g) 27 009    h) 18 207
   i) 33 025    j) 54 916    k) 30 900    l) 51 842

2) a) 385 091    b) 257 069    c) 885 819
   d) 463 795    e) 488 288    f) 787 314
   g) 366 562    h) 442 601    i) 460 725
   j) 451 004    k) 622 240    l) 620 183

### Challenge
Yes, Finlay is correct: 2 000 000 − 1 000 000 = 1 000 000;
700 000 − 400 000 = 300 000;
85 − 70 = 15 (other explanations are acceptable).

## 3.2 Adding and subtracting a string of numbers (p.36)

### Before we start
1860, 3177 and 1003

### Questions
1) a) 10 860    b) 20 630    c) 27 499    d) 3890
   e) 30 190    f) 37 104    g) 7580

2) a) 13 095 (subtract 1000 altogether)
   b) 22 426 (subtract 1300 altogether)
   c) 6000 (subtract 14 000 altogether)
   d) 15 060 (subtract 2040 altogether)
   e) 43 000 (subtract 5080 altogether)

3) a) 1275 + 2**425** + 803 = 4503
   b) 9**673** − 360 − 240 = 9073

### Challenge
There are six possible combinations:
1225 and 8775    2125 and 7875
2215 and 7785    1775 and 8225
7175 and 2825    7715 and 2285

## 3.3 Using place value partitioning to add and subtract (p.38)

### Before we start
a) Four × 10 000 counters, eight × 1000 counters, two × 100 counters, four × 10 counters

b) Six × 10 000 counters, five × 100 counters, five × 1 counters

c) Two × 10 000 counters, three × 1000 counters, one × 10 counter, seven × 1 counters

d) Five × 100 000 counters, one × 1 counter

### Questions
1) a) 159 978    b) 694 799    c) 406 388
   d) 51 372    e) 136 407    f) 54 431

2) a) 89 307    b) 179 759    c) 546 742
   d) 40 386    e) 117 209    f) 435 799

3) a) 831 559    b) 898 863    c) 602 418
   d) 715 068    e) 538 075    f) 16 500

### Challenge
600 000 and 400 000    160 100 and 839 900
215 000 and 785 000    421 060 and 578 940

## 3.4 Adding whole numbers using standard algorithms (p.40)

### Before we start
1) a) The ten carried over from the ones column has not been included. The answer should be 5935.

   b) Ten thousand is missing from the answer, which should read 12 082.

   c) Zero is needed as a placeholder in the ones column. The answer should read 6440.

### Questions
1) a) 114 357    b) 126 820    c) 223 140
   d) 762 246    e) 682 804    f) 866 264
   g) 882 625    h) 1 209 317

2) a) 84 556    b) 161 464    c) 72 996
   d) 166 008    e) 150 668    f) 210 822

3) a) Various answers are possible. The digits 0–9 should be used once only in each algorithm.

   i) should include carrying a thousand into the ten thousands column

   ii) should include carrying a ten into the hundreds column

   iii) should include carrying over a ten, a hundred and a thousand

   b) Various answers are possible.

   c) Various answers are possible.

### Challenge
Should be as follows with the bold digits in squares

```
  5 1 1 4 2 9          2 0 6 4 1 3
+ 2 8 5 0 7 6        + 1 0 5 8 2 8
  7 9 6 5 0 5          3 1 2 2 4 1
```

## 3.5 Subtracting whole numbers using standard algorithms (p.42)

### Before we start

a)
```
  3 9 12
  4 0 3 2
− 2 3 5 6
  1 6 7 6
```

b)
```
  2 9
  8 3 0 7
− 1 1 4 8
  7 1 5 9
```

c)
```
  4 11 9
  5 2 0 1
− 3 8 6 3
  1 3 3 8
```

### Questions
1) a) 12 785    b) 206 215    c) 315 261    d) 220 491
   e) 159 085    f) 645 594    g) 403 512    h) 342 963

2) a) 29 494    b) 55 665    c) 83 771    d) 378 225
   e) 292 255    f) 608 215    g) 85 695    h) 233 888

3) a) Answers will vary. The digits 0–9 should be used once only in each algorithm.

   i) should include the need to exchange 1 thousand for 10 hundreds

   ii) should include the need to exchange 1 ten thousand for 10 one thousands

iii) should include the need to exchange 1 ten thousand for 10 thousands **and** 1 ten for 10 ones

b) Answers will vary.  c) Answers will vary.

## Challenge

| 8 | 1 | 7 | 6 | 5 | **8** |
|---|---|---|---|---|---|
| 2 | **0** | 9 | 5 | 8 | 3 |
| 6 | 0 | 8 | 0 | **7** | 5 |

| 6 | **2** | 0 | **4** | 3 | 1 |
|---|---|---|---|---|---|
| **1** | 7 | 5 | 2 | 8 | 5 |
| 4 | 4 | 5 | 1 | **4** | **6** |

## 3.6 Mental and written calculation strategies (p.44)

### Before we start

17 500

### Questions

1) a) 500 000   b) 493 998   c) 830 514
   d) 189 999   e) 954 364   f) 13 036

2) Second row from the bottom from left to right: 7315, 4919, 3591, 3400
   Middle row: 2396, 1328, 191
   Second row from the top: 1068, 1137
   Top brick: 69

### Challenge

Answers will vary.

## 3.7 Representing word problems (p.46)

### Before we start

Isla needs to add 9486 and 13 566, then subtract her answer from 24 000. The answer is 948 empty seats.

### Questions

1) Ben Nevis received **384 278** visitors.
   The number sentence 247 139 − 137 139 = **384 278** should be correctly represented as a bar model and empty number line.

2) a) These attractions received a total of **1 731 646** visitors.
   The number sentence 521 710 + 642 677 + 567 259 = 1 731 646 should be correctly represented as a bar model and empty number line.
   The addition algorithm would be an appropriate choice of method.

   b) Urquhart Castle received **380 152** visitors.
   The number sentence 567 259 − 187 107 should be correctly represented as a bar model and empty number line.
   The subtraction algorithm is an acceptable method here. Some pupils may use place value partitioning, interpreting 560 000 − 180 000 as 38 × 100 000.

3) 2 063 709 is **1 063 709** more than 1 000 000.
   The number sentence 2 063 709 − 1 000 000 = 1 063 709 should be correctly represented as a bar model and empty number line. Jottings that show the pupil understands that **2** 063 709 is one million more than **1** 063 709 are appropriate.

4) The National Museum of Scotland, Edinburgh, had **810 242** more visitors than the Riverside Museum, Glasgow.
   The number sentence 2 165 601 − 1 355 359 should be correctly represented as a bar model and empty number line. The subtraction algorithm would be an appropriate choice of method.

### Challenge

Answers will vary.

## 3.8 Solving multi-step word problems (p.48)

### Before we start

a) 555 380 = 598 830 − 43 450
b) 482 005 − (207 932 + 148 270) = 125 803

### Questions

1) a) £182 000   b) £10 634

2) 300 731 points

3) 95 small creatures were treated

4) 10 849 km

### Challenge

1) a) fifty million kilometres   50 000 000 km
   b) one hundred and seventy million kilometres   170 000 000 km

2) six hundred and twenty-eight million, nine hundred thousand kilometres 628 900 000 km

## 3.9 Adding whole numbers and decimal fractions (p.50)

### Before we start

Finlay has disregarded the decimal point and added 37 and 45. The correct calculation is 37 + 4·5, giving a correct answer of 41·5.

### Questions

1) a) 65·89   b) 46·71   c) 31·02   d) 94·55
   e) 60·64   f) 100·92   g) 81·31   h) 412·13
   i) 560·85   j) 581·22   k) 266·91   l) 950·44

2) a) 8·07   b) 10·15   c) 14·79   d) 6·14
   e) 13·51   f) 18·75   g) 8·08   h) 17·04
   i) 14·58   j) 13·98   k) 7·96   l) 8·93

3) a) 23·79   b) 31·76   c) 87·57
   d) 95·87   e) 60·89   f) 57·85

4) a) 27·74   b) 14·63   c) 55·98
   d) 316·35   e) 180·2   f) 28·97

### Challenge

a) 2670 and 0·88      2600 and 70·88
   1330·87 and 1340·01      1000·45 and 1670·43

b) Any pair/pairs of decimal fractions that total 2670·88.

## 3.10 Adding decimal fractions using standard written algorithms (p.52)

### Before we start

Any three pairs of decimal fractions that add to 1.

### Questions

1) a) 42·37   b) 80·95   c) 86·15   d) 188·61
   e) 418·25   f) 271·07   g) 202·12   h) 490·21
   i) 395·82   j) 907·43   k) 1318·05   l) 673·4

2) a) 22·69   b) 24·31   c) 64·67   d) 212·8(0)

3) a) More than one solution is possible. Answers should be addition algorithms comprising two decimal fractions made from the digits: 1, 3, 4, 5, 6, 7, 8, 9. Solutions should also meet the following criteria:
   i) answer has 5 in the hundredths column
   ii) answer is > 100
   iii) only need to carry one tenth over from the hundredths column
   iv) three carrying figures

   b) Adding any two decimal fractions with two decimal places, made from the digits above, will give an answer greater than 50. The answer in this instance is therefore 'no'. However, it is possible to create two decimal fractions with three decimal places to give an answer <50.

### Challenge

| 1 | 4 | **6** | · | 2 | 8 |
|---|---|---|---|---|---|
| + | **2** | 0 | 4 | · | **5** | 3 |
| 3 | 5 | 0 | · | 8 | 1 |

| 5 | 2 | 9 | · | 8 | 7 |
|---|---|---|---|---|---|
| + | 3 | 8 | 5 | · | **1** | 6 |
| **9** | **1** | **5** | · | 0 | 3 |

| 4 | **0** | 2 | · | 8 | 5 |
|---|---|---|---|---|---|
| + | 4 | 2 | **0** | · | 3 | 5 |
| 8 | 2 | 3 | · | 2 | **0** |

Zero is needed in the first two answers, as they are acting as placeholders. The zero in the third answer is not required, as 823·20 means the same as 823·2.

## 3.11 Subtracting decimal fractions (p.54)

**Before we start**

a) 32·3      b) 48·4      c) 41      d) 30·1

**Questions**

1)
a) 4·57    b) 6·05    c) 5·08    d) 11·24
e) 9·42    f) 4·46    g) 6·15    h) 6·24
i) 0·81    j) 160·26    k) 173·64    l) 200·9(0)

2)
a) 12·54    b) 40·72    c) 68·04    d) 41·83
e) 50·68    f) 27·49    g) 25·77    h) 344·45
i) 613·97    j) 596·02    k) 907·71    l) 1000·59

3) The following calculations are possible. Pupils should explain the strategy they used to arrive at the answers shown.

4·29 – 1·03 = 3·26      4·29 – 3 = 1·29
4·29 – 2·65 = 1·64      3 – 2·65 = 0·35
3 – 1·03 = 1·97      2·65 – 1·03 = 1·62

**Challenge**

a) 21·12

b) Clues made up of two decimal fractions with a difference of: 47·74, 39·93 and 80·08

## 3.12 Subtracting decimal fractions using standard algorithms (p.56)

**Before we start**

```
  9 1 3  8 2 0        6 2 1  0 8 7
– 1 4 5  2 3 6      – 4 1 1  5 9 1
  7 6 8  5 8 4        2 0 9  4 9 6
```

**Questions**

1)
a) 22·63    b) 16·56    c) 64·71    d) (0)6·97
e) 18·29    f) 24·45    g) (0)5·62    h) (0)2·59
i) 33·54    j) 28·78    k) (0)1·69    l) (0)0·77
m) 14·7(0)    n) 40·19    o) 50·06

2)
a) Compare answers to question 1 with a partner.

b) Not all of the zeros are needed because not all are place holders. For example, 06·97 means the same as 6·97.

c) The following answers have zero as a place holder:
0·77 (only the zero in the ones column is needed because 00·77 means the same as 0·77)
40·19 (zero is needed to keep the 4 tens in place)
50·06 (the zeros are needed to keep the 5 tens and 6 hundredths in place)

3) 80·15 – 39·96 could be solved by rounding and adjusting →
80·15 – 40 + 0·04 = 40·19

80·15 – 39·96 could also be solved by transforming both numbers → 80·19 – 40 = 40·19

67·75 – 17·69 could be solved by partitioning → 67 – 17 = 50 and 0·75 – 0·69 = 0·06 → 50 + 0·06 = 50·06

**Challenge**

a) Pairs with a difference of 13·64 are:
21 and 7·36    31·1 and 17·46    30·46 and 16·82
62·75 and 49·11    72·22 and 58·58

b) Any two decimal fractions that have a difference of 13·64

## 3.13 Adding and subtracting decimal fractions (p.58)

**Before we start**

The most efficient strategy in the first example is round and adjust.
3·25 + 16·99 → 3·24 + 17 = 17 + 3·24 = 20·24
An algorithm is appropriate in the second example 73·29 – 7·42 = 65·87

**Questions**

1)
a) 16·44    b) 28·02    c) 51·35    d) 53·78
e) 102·15    f) 28·79    g) 159·78    h) 500·77
i) 615·21    j) 20·04    k) 40·04

2)
a) 1·45    b) 0·58    c) 6·07    d) 12·28
e) 12·61    f) 28·59    g) 55·36    h) 31·83
i) 40·79    j) 119·8    k) 145·35    l) 743·76

3)
b) + 0·5    c) + 0·06    d) + 0·29
e) + 0·7    f) + 2·1    g) – 0·35
h) – 1·1    i) – 6·06    j) – 2·3

**Challenge**

a) False 522·18 ≠ 521·08      b) False 24·13 ≠ 25·13

## 3.14 Representing word problems involving decimal fractions (p.60)

**Before we start**

35·8 means £35·80    Finlay still needs to save £5·65

**Questions**

1) Problem correctly represented as a bar mode and empty number line. Correct mental or written strategy used to give an answer of 5·75 m.

2) Problem correctly represented as a bar mode and empty number line. Correct mental or written strategy used to give an answer of 23·84 seconds.

3) Problem correctly represented as a bar mode and empty number line. Correct mental or written strategy used to give an answer of £72·83.

4) Problem correctly represented as a bar mode and empty number line. Correct mental or written strategy used to give an answer of 70·25 kg.

5) Problem correctly represented as a bar mode and empty number line. Correct mental or written strategy used to give an answer of 932·47 kg.

**Challenge**

Answers will vary.

## 3.15 Multi-step word problems (p.62)

**Before we start**

The Turkish athlete took less than double the time to run double the distance (19·76 seconds compared to 19·9 seconds).

**Questions**

1) 921·81 m

2) 23·12 m

3) Coat £129·97      Shoes £50·70      Bag £35·25

4) Crate D weighs 53·85 kg

**Challenge**

a) The combined mass of the creatures is 138 787·905 kg

b) The hippo weighs 3749·625 kg more than the hamster.

## 4 Number – multiplication and division

## 4.1 Multiplication and division facts for 7 (p.64)

**Before we start**

Several strategies are possible, for instance 30 × 4 = 120 so we need to subtract 2 × 4.
So 112 = 28 × 4 and 112 ÷ 4 = 28

**Questions**

1)
a) 28    b) 56    c) 42    d) 63
e) 49    f) 35    g) 77

2)
a) 3    b) 56    c) 21    d) 5
e) 9    f) 2    g) 7    h) 6
i) 10    j) 8    k) 28    l) 9

3)
a) Check that multiplication triangles have been completed correctly.
b) Answers will vary.      c) Answers will vary.

## Challenge

Answers will vary – here are some examples:

$11 \times 7 = (11 \times 10) - (11 \times 3) = 110 - 33 = 77$
$14 \times 7 = (14 \times 10) - (14 \times 3) = 140 - 42 = 98$
$15 \times 7 = (15 \times 10) - (15 \times 3) = 150 - 45 = 105$
$20 \times 7 = (20 \times 10) - (20 \times 3) = 200 - 60 = 140$

## 4.2 Recalling multiplication and division facts for 8 (p.66)

### Before we start

$5 \times 14 = 5 \times 7 + 5 \times 7 = 35 + 35 = 70$

### Questions

1) a) 40    b) 56    c) 16    d) 32

2) Check that the pyramid has been completed correctly.

3) Answers will vary.

### Challenge

96    120    144    160

Method works since 10 lots of a number minus 2 lots will give 8 lots. Nuria could have used the double 3 times method.

## 4.3 Multiplication with decimal fractions (p.68)

### Before we start

a) 2300 g    b) 5200 g    c) 16 100 g
d) 900 g    e) 468 000 g    f) 1100 g

### Questions

1) a) 22·8    b) 63    c) 1846    d) 14 550
   e) 502    f) 8924·9    g) 30    h) 1739

2) a) **True**    b) **False**    c) **False**    d) **False**
   e) **True**    f) **False**    g) **False**    h) **True**

3) a) 10    b) 100    c) 1000
   d) 100    e) 10    f) 100

### Challenge

£319 for 1000 pencils.

## 4.4 Division with decimal fractions (p.70)

### Before we start

£694    £69·40

### Questions

1) a) 5·2    b) 0·06    c) 0·412
   d) 8·2    e) 0·99    f) 1·25

2)

| Millimetres | Centimetres | Metres |
|---|---|---|
| 1560 | 156 | 1·56 |
| 67 240 | 6724 | 67·24 |
| 9890 | 989 | 9·89 |
| 4 930 000 | 493 000 | 4930 |
| 114 560 | 11 456 | 114·56 |

3) a) £8·92 per metre   b) £7·99 per metre   c) £10·01 per metre

### Challenge

1400

## 4.5 Solving multiplication problems (p.72)

### Before we start

£65 300 to multiply by 100 we can add two zeros.

### Questions

1) a) 28 674    b) 36 496    c) 37 460
   d) 11 238    e) 30 058    f) 33 255

2) a) 22 449    b) 34 424    c) 31 734
   d) 14 640    e) 33 859    f) 17 442

3) a) £5726    b) £20 041    c) £25 767

### Challenge

1) **Yes** Isla did order enough since $1075 \times 3 = 3225 > 3200$

2) **Yes** Isla did order enough since $1685 \times 4 = 6740$ and $3200 \times 2 = 6400$

## 4.6 Solving multiplication problems involving decimal fractions (p.74)

### Before we start

7 tenths
9 hundredths
8·89

### Questions

1) b) 7·62    c) 7·2    d) 26·01    e) 8·97

2) b) 46·9    c) 28·35    d) 14·32    e) 32·76

3) a) £2·67    b) £8·20    c) £4·92    d) £19·60
   e) £23·82

### Challenge

Answers will vary.

## 4.7 Solving division problems using place value (p.76)

### Before we start

14 days    $10 \times 6 = 60$ and $4 \times 6 = 24$ so $14 \times 6 = 60 + 24 = 84$

### Questions

1) Check that the correct grid method has been used to find the following answers:
   b) 139    c) 131    d) 63    e) 72    f) 122

2) a) $275 \div 5 = 55$    b) $378 \div 3 = 126$    c) $412 \div 4 = 103$
   d) $552 \div 6 = 92$    e) $670 \div 5 = 134$

3) a)

| 387 | 702 | 441 | 549 |
|---|---|---|---|
| 43 | 78 | 49 | 61 |

   b)

| 216 | 344 | 536 | 416 |
|---|---|---|---|
| 27 | 43 | 67 | 52 |

### Challenge

a) $824 \div 4 = 206$      b) $7235 \div 5 = 1447$

c) $3954 \div 6 = 659$      d) $8435 \div 7 = 1205$

## 4.8 Solving multiplication problems using proportional adjustment (p.78)

### Before we start

There are several strategies available including
$7 \times 18 = (7 \times 10) + (7 \times 8) = 70 + 56 = 126$
$7 \times 18 = (7 \times 20) - (7 \times 2) = 140 - 14 = 126$
$7 \times 18 = (7 \times 9) + 7 \times (9 = 63) + 63 = 126$

### Questions

1) b) Correct    c) Correct    d) Correct    e) Correct

2) a) $4 \times 22 = 8 \times 11 = 88$    b) $3 \times 16 = 6 \times 8 = 48$
   c) $5 \times 24 = 10 \times 12 = 120$    d) $18 \times 4 = 9 \times 8 = 72$
   e) $28 \times 5 = 14 \times 10 = 140$    f) $14 \times 3 = 7 \times 6 = 42$

3) a) $12 \times 7 = 4 \times 21 = £84$
   b) $5 \times 42 = 10 \times 21 = 210$ rows
   c) $3 \times 24 = 6 \times 12 = £72$

### Challenge

Answers will vary.

## 4.9 Solving division problems (p.80)

### Before we start

Arrays should show
$1 \times 48$    $2 \times 24$    $3 \times 16$    $4 \times 12$    $6 \times 8$

## Questions

1) a) $54 \div 3 = (60 \div 3) - (6 \div 3) = 18$
   b) $95 \div 5 = \boxed{(100 \div 5) - (5 \div 5)} = \boxed{19}$
   c) $96 \div 4 = \boxed{(100 \div 4) - (4 \div 4)} = \boxed{24}$
   d) $76 \div 4 = \boxed{(80 \div 4) - (4 \div 4)} = \boxed{19}$
   e) $144 \div 3 = \boxed{(150 \div 3) - (6 \div 3)} = \boxed{48}$

2) 24

3) a) 38    b) 19

## Challenge

$192 \div 4 = (200 \div 4) - (8 \div 4) = 50 - 2 = £48$

## 4.10 Solving multiplication problems (p.82)

### Before we start

Her answer is wrong as any multiple of 5 ends in 5 or 0.
Answer should be 140.

### Questions

1) a)

| × | 10 | 4 |
|---|---|---|
| 20 | 200 | 80 |
| 3 | 30 | 12 |
|  | 230 | 92 |

$14 \times 23 = 322$

b)

| × | 30 | 1 |
|---|---|---|
| 40 | 1200 | 40 |
| 5 | 150 | 5 |
|  | 1350 | 45 |

$31 \times 45 = 1395$

c)

| × | 70 | 3 |
|---|---|---|
| 10 | 700 | 30 |
| 6 | 420 | 18 |
|  | 1120 | 48 |

$73 \times 16 = 1168$

d)

| × | 50 | 2 |
|---|---|---|
| 20 | 1000 | 40 |
| 8 | 400 | 16 |
|  | 1400 | 56 |

$52 \times 28 = 1456$

2) a) $28 \times 33$

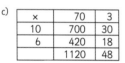

|   |   | 2 | 8 |
|---|---|---|---|
| × |   | 3 | 3 |
|   |   | 8 | 4 |
| 8 | 4 | 0 |   |
| 9 | 2 | 4 |   |

b) $19 \times 23$

|   |   | 1 | 9 |
|---|---|---|---|
| × |   | 2 | 3 |
|   |   | 5 | 7 |
| 3 | 8 | 0 |   |
| 4 | 3 | 7 |   |

c) $21 \times 42$

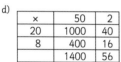

|   |   | 2 | 1 |
|---|---|---|---|
| × |   | 4 | 2 |
|   |   | 4 | 2 |
| 8 | 4 | 0 |   |
| 8 | 8 | 2 |   |

d) $27 \times 19$

|   |   | 2 | 7 |
|---|---|---|---|
| × |   | 1 | 9 |
|   | 2 | 4 | 3 |
|   | 2 | 7 | 0 |
|   | 5 | 1 | 3 |

3) a) 1148 ml    b) £420    c) 945 chairs

## Challenge

Answers will vary.

## 4.11 Solving division problems using an algorithm (p.84)

### Before we start

Various strategies are possible
$87 \div 3 = 90 \div 3 - 3 \div 3 = 30 - 1 = 29$
$87 \div 3 = 60 \div 3 + 27 \div 3 = 20 + 9 = 29$
$87 = 30 + 30 + 27$ So $87 \div 3 = 10 + 10 + 9 = 29$

### Questions

1) a) 19    b) 17    c) 26    d) 12

2) a) 29    b) 18    c) 26    d) 13
   e) 13    f) 116

3) a) 42    b) 28    c) 24

## Challenge

Answers will vary.

## 4.12 Solving division problems involving decimal fractions (p.86)

### Before we start

Answers will vary.

### Questions

1) a) 0·31    b) 2·19    c) 0·13    d) 3·09

2) a) 0·95    b) 0·64    c) 0·96
   d) 0·63    e) 0·66    f) 0·71

3) a) £0·47    b) £0·98    c) £0·32
   d) £1·26    e) £0·87    f) £1·56

## Challenge

Neither is correct. $4·16 \div 4 = 1·04$

Possible strategies could include – written method, halving and halving again, partitioning by place value, reversing.

## 4.13 Solving multi-step problems using the order of operations (p.88)

### Before we start

$38 - 4 \times 8 = 6$ so £6.

### Questions

1) a) $(4 \times 3) - 6 = 6$    b) $21 - (2 \times 9) + 5 = 8$
   c) $(9 \times 2) + (8 \div 4) = 20$    d) $(42 \div 7) - 3 + 16 = 19$
   e) $4 + (27 \div 3) - 4 = 9$    f) $(25 \div 5) + (18 \div 6) = 8$
   g) $(7 \times 4) + (9 \times 2) - 5 = 41$    h) $16 - (6 \div 2) + (5 \times 4) = 33$

2) a) $6 + 14 \div 2 = (6 + 14) \div 2$ Incorrect. $6 + (14 \div 2) = 13$
   b) $5 \times 6 - 2 \times 3 = (5 \times 6) - (2 \times 3)$ Correct.
   c) $14 - 4 \div 2 + 8 \times 4 = (14 - 4) \div 2 + (8 \times 4)$ Incorrect.
      $14 - (4 \div 2) + (8 \times 4) = 44$
   d) $48 \div 8 - 4 + 5 \times 2 = (48 \div 8) - (4 + 5) \times 2$ Incorrect.
      $(48 \div 8) - 4 + (5 \times 2) = 12$

3) a) £9    b) £39    c) £24

## Challenge

1) a) $33 \times 3 = 99$    b) $99 \times 9 = 891$    c) $41 \times 5 = 205$

2) a) $111 \times 4 = 444$    b) $555 \times 5 = 2775$    c) $301 \times 3 = 903$

# 5 Multiples, factors and primes

## 5.1 Using knowledge of multiples and factors to work out divisibility rules (p.90)

### Before we start

Nuria is wrong. Factors of 12 are 1, 2, 3, 4, 6, 12. She has forgotten 1 and 12.

### Questions

1) a) 75  890  47 855  7540  890 635  182 780  748 905
   b) The last digit of a number must be 5 or 0 to be divisible by 5.

2) The last digit of a number must be 0 to be divisible by 10. Multiples of 10 always end in 0.

3) a) **False**    b) **True**    c) **True**    d) **False**
   e) **False**    f) **True**    g) **False**    h) **True**

## Challenge

a) 126  1026  312  216  684  1524  372  528

b) A number is divisible by 6 if the sum of the digits in a number can be divided by 3 and the number is even, or can also be divided by 2.

c) The rule for divisibility by 8 is to look at only the last three digits. If that 3-digit number is divisible by 8 then the whole number is.

## 5.2 Using knowledge of multiples and factors to solve problems (p.92)

**Before we start**

List any three multiples of 10.

We notice that all multiples of 10 are multiples of 2 and 5 also.

**Questions**

1) a) No, because the last digit is an odd number we know it can't be a multiple of 6.

   b) 53 boxes    c) 27 boxes

2) $1 \times 64$, $2 \times 32$, $4 \times 16$, $8 \times 8$

3)

| Problem | Nuria's answer | Amman's answer | Who is correct? |
|---------|---------------|----------------|-----------------|
| $73 \times 5$ | 438 | 365 | Amman. Multiples of 5 always end in a 5 or a 0. |
| $155 \div 2$ | 75 | 77.5 | Amman. An odd number can't be a multiple of 2. |
| $26 \times 8$ | 208 | 64 | Nuria. 26 isn't a factor of 64. |
| $36 \times 9$ | 324 | 360 | Nuria. If 36 is a factor of 360, the other factor must be 10. |

**Challenge**

Isla's dad must be 35.

## 6 Fractions, decimal fractions and percentages

### 6.1 Converting fractions (p.94)

**Before we start**

$\frac{7}{10}$    $\frac{1}{4}$    $\frac{7}{8}$    $\frac{7}{10}$

**Questions**

1) a) $\frac{5}{3} = 2\frac{2}{3}$    b) $\frac{7}{2} = 3\frac{1}{2}$    c) $\frac{9}{4} = 2\frac{1}{4}$

   d) $\frac{8}{5} = 1\frac{3}{5}$    e) $\frac{14}{6} = 2\frac{2}{6}$ (or $2\frac{1}{3}$)

2) a) $1\frac{3}{4} = \frac{7}{4}$    b) $1\frac{1}{2} = \frac{3}{2}$    c) $2\frac{1}{3} = \frac{7}{3}$

   d) $2\frac{4}{5} = \frac{14}{5}$    e) $1\frac{5}{6} = \frac{11}{6}$

3) a) Amman: $4\frac{1}{2} = \frac{9}{2}$

   b) Isla: $2\frac{3}{4} = \frac{11}{4}$

   c) Finlay: $2\frac{8}{10}$ (or $2\frac{4}{5}$) $= \frac{28}{10}$ (or $\frac{14}{5}$)

   d) Nuria: $3\frac{4}{24}$ (or $3\frac{1}{6}$) $= \frac{76}{24}$ (or $\frac{19}{6}$)

**Challenge**

1) $\dfrac{19}{6}$  $\dfrac{16}{5}$  $\dfrac{10}{3}$  $\dfrac{51}{15}$  $\dfrac{35}{10}$

2) $\dfrac{29}{12}$  $\dfrac{89}{24}$  $\dfrac{33}{20}$  or  $\dfrac{34}{20}$

### 6.2 Comparing and ordering fractions (p.97)

**Before we start**

$\frac{16}{24}$  $\frac{14}{24}$  $\frac{18}{24}$  $\frac{15}{24}$

So smallest to largest is $\frac{7}{12}$  $\frac{5}{8}$  $\frac{2}{3}$  $\frac{3}{4}$

**Questions**

1) a) Nuria: $\frac{2}{3} = \frac{8}{12}$, Isla: $\frac{3}{4} = \frac{9}{12}$. Isla has read more.

   b) Isla: $\frac{3}{8} = \frac{9}{24}$, Amman: $\frac{5}{12} = \frac{10}{24}$. Amman uses more of his clay.

   c) Amman: $\frac{3}{5} = \frac{21}{35}$, Finlay: $\frac{4}{7} = \frac{20}{35}$. Amman has more juice.

2) b) $\frac{7}{2} = \frac{14}{4}$. $\frac{7}{2}$ is less than $\frac{15}{4}$

   c) $\frac{5}{3} = \frac{20}{12}$, $\frac{7}{4} = \frac{21}{12}$. $\frac{5}{3}$ is less than $\frac{7}{4}$

   d) $\frac{12}{5} = \frac{48}{20}$, $\frac{9}{4} = \frac{45}{20}$. $\frac{12}{5}$ is larger than $\frac{9}{4}$

   e) $\frac{11}{3} = \frac{55}{15}$, $\frac{17}{5} = \frac{51}{15}$. $\frac{11}{3}$ is larger than $\frac{17}{5}$

**Challenge**

The children can write each of their scores as a simplified fraction:

Isla: $\frac{15}{25} = \frac{3}{5}$    Finlay: $\frac{28}{56} = \frac{1}{2}$

Nuria: $\frac{22}{33} = \frac{2}{3}$    Amman: $\frac{24}{32} = \frac{3}{4}$

They can then find a common equivalent:

Isla: $\frac{36}{60}$  3rd    Finlay: $\frac{30}{60}$  4th

Nuria: $\frac{40}{60}$  2nd    Amman: $\frac{45}{60}$  1st

### 6.3 Simplifying fractions (p.100)

**Before we start**

$\frac{4}{5}$ and $\frac{1}{3}$

**Questions**

1)

a)

nine fifteenths    $\frac{9}{15} \overset{\div 3}{\underset{\div 3}{=}} \frac{3}{5}$

b)

55 eightieths    $\frac{55}{80} \overset{\div 5}{\underset{\div 5}{=}} \frac{11}{16}$

c)

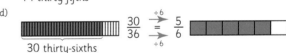

14 thirty-fifths    $\frac{14}{35} \overset{\div 7}{\underset{\div 7}{\rightarrow}} \frac{2}{5}$

d)

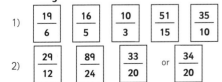

30 thirty-sixths    $\frac{30}{36} \overset{\div 6}{\underset{\div 6}{\rightarrow}} \frac{5}{6}$

2) b) $\frac{3}{5}$    c) $\frac{2}{3}$    d) $\frac{3}{5}$

**Challenge**

a) i) $\frac{3}{4}$    ii) $\frac{2}{3}$    iii) $\frac{7}{10}$    iv) $\frac{3}{5}$

b) ii) has the most common factors.

### 6.4 Adding and subtracting fractions (p.102)

**Before we start**

Many possibilities such as $\frac{4}{6}$ $\frac{6}{9}$ $\frac{8}{12}$ $\frac{20}{30}$ and $\frac{6}{8}$ $\frac{9}{12}$ $\frac{12}{16}$ $\frac{30}{40}$

**Questions**

1) a) $\frac{1}{3} + \frac{1}{6} = \frac{3}{6}$ (or $\frac{1}{2}$)    b) $\frac{3}{4} + \frac{3}{8} = \frac{9}{8}$ (or $1\frac{1}{8}$)

   c) $\frac{7}{10} - \frac{2}{5} = \frac{3}{10}$    d) $\frac{10}{6} - \frac{9}{12} = \frac{11}{12}$

2) a) $\frac{11}{10}$ or $1\frac{1}{10}$    b) $\frac{5}{8}$    c) $\frac{17}{12}$ or $1\frac{5}{12}$

d) $1\frac{4}{6}$ or $1\frac{2}{3}$

3) a) $3\frac{2}{12}$ miles (or $3\frac{1}{6}$ miles)

b) $1\frac{5}{10}$ kg (or $1\frac{1}{2}$ kg)

c) $2\frac{3}{6}$ miles (or $2\frac{1}{2}$ miles)

## Challenge

a) $\frac{1}{6}$ of the children have brown eyes so 55 children have brown eyes.

b) $\frac{11}{20}$ of the sweets are left in the bag.

## 6.5 Converting decimal fractions to fractions (p.104)

**Before we start**

$\frac{3}{10}$    $\frac{55}{100} = \frac{11}{20}$    $\frac{79}{100}$    $\frac{8}{10} = \frac{4}{5}$

### Questions

1) 0·65 = 13 twentieths        0·67 = two thirds
0·7 = seven tenths        0·75 = three quarters
0·72 = 18 twenty-fifths        0·6 = three fifths

2) a) $0·75 = \frac{3}{4}$    b) $0·35 = \frac{7}{20}$    c) $0·12 = \frac{3}{25}$

## Challenge

a) $\frac{22}{25}$    b) $1\frac{9}{20}$    c) $3\frac{9}{10}$    d) $5\frac{18}{25}$    e) $4\frac{3}{5}$

## 6.6 Calculating a fraction of a fraction (p.106)

**Before we start**

Each third is 120 people so 240 support the red team.

### Questions

1) a) $\frac{1}{12}$    b) $\frac{3}{8}$    c) $\frac{1}{3}$    d) $\frac{1}{6}$

2) a) $\frac{1}{8}$ of a bag of sweets    b) $\frac{1}{9}$ of the whole cake

c) $\frac{1}{6}$ of a bottle

## Challenge

a) Trainers: $\frac{3}{10}$ of his birthday money

b) Football top: $\frac{1}{5}$ of his birthday money

c) Headphones: $\frac{1}{10}$ of his birthday money

## 6.7 Dividing a fraction by a whole number (p.108)

**Before we start**

Answers will vary but must contain one quarter and out of 32.

### Questions

1) a) $\frac{1}{4} \div 3 = \frac{1}{12}$    b) $\frac{1}{5} \div 4 = \frac{1}{20}$    c) $\frac{1}{3} \div 4 = \frac{1}{12}$

2) a) $\frac{3}{4} \div 4 = \frac{3}{16}$    b) $\frac{3}{5} \div 2 = \frac{3}{10}$    c) $\frac{5}{6} \div 3 = \frac{5}{18}$

## Challenge

Amman - 13 muffins        Nuria - 14 pancakes
Finlay - 32 scones        Isla - 34 cupcakes

## 6.8 Dividing a whole number by a fraction (p.110)

**Before we start**

Answers will vary but must contain three fifths and out of 200.

## Questions

1) a) 24 portions of steak    b) 40 portions of lasagne
c) 48 portions of pizza    d) 40 portions of cheesecake
e) 72 portions of gateaux

2) a) 21    b) 20    c) 8

## Challenge

a) 16 jugs, 15 large glasses and 12 small glasses.

b) $12\frac{1}{2}$ large glasses

## 6.9 Equivalents: fractions, decimals and percentages (p.112)

**Before we start**

$\frac{1}{10}$    $\frac{1}{4}$    $\frac{76}{100} = \frac{19}{25}$    $2\frac{8}{10} = 2\frac{4}{5}$

### Questions

1) a) 0·9 and 90%        b) 0·4 and 40%
c) 0·625 and 62·5%        d) 0·35 and 35%

2) a) $0·6 = \frac{6}{10} = \frac{3}{5} = 60\%$    b) $0·85 = \frac{85}{100} = \frac{17}{20} = 85\%$

c) $0·34 = \frac{34}{100} = \frac{17}{50} = 34\%$

## Challenge

a) None of these fractions can be directly converted into tenths or hundredths.

b) $\frac{1}{3} = 0·33$        $\frac{1}{6} = 0·17$        $\frac{1}{8} = 0.125$

$\frac{1}{9} = 0·11$        $\frac{1}{12} = 0·08$        $\frac{1}{15} = 0·07$

## 6.10 Solving fraction, decimal and percentage problems (p.114)

**Before we start**

$\frac{85}{100}$ or $\frac{17}{20}$ or 85%    0.6 or 60%    $\frac{24}{100}$ $3\frac{6}{25}$ or 24%

### Questions

1) a) 3500    b) 5700    c) 6660

2) Answers will vary.

3) a) 860 km    b) 2925 likes    c) 26·25 litres (or $26\frac{1}{4}$ litres)

## Challenge

Answers will vary.

# 7 Money

## 7.1 Money problems using the four operations (p.116)

**Before we start**
a) £27·50        b) £20 note, £5 note, £2 coin and 50p

### Questions

1) a) £35·20        b) Gamers Direct

2) a) £8·50        b) A book of four return tickets

3) a) LANDSCAPE LARRY £40    b) GARDENING GURU £8·00

## Challenge

a) DVD from TV'S "R" US, TV from ELECTRIC AVENUE, DVD boxset from PRICE PERFECT

b) £96·50        c) £103·20

## 7.2 Budgeting (p.118)

**Before we start**

10 weeks

## Questions

1) a) £375    b) £287    c) Yes

2)

|  | Credit | Debit |
|---|---|---|
| Income | £55·00 |  |
| Comic (£1·50 per week) |  | £6·00 |
| Cinema (£5·00 per week) |  | £20·00 |
|  |  |  |
| Totals | **£55·00** | **£26·00** |

3) £550

### Challenge

a) £565    b) £365

## 7.3 Profit and loss (p.121)

### Before we start

1) £550    2) £4·50 each

### Questions

1) £30 loss

2) £8885

3) £30

### Challenge

50 tickets at £10·50 each (or other acceptable answers)

## 7.4 Discounts (p.123)

### Before we start

1) GADGET GARAGE    2) £231

### Questions

1) £54

2) 'buy one get one half price'

### Challenge

£10·35 at the SUPERMARKET as it would be £11·15 at FOOD & STUFF.

## 7.5 Hire purchase (p.125)

### Before we start

Answers will vary.

### Questions

1) Answers will vary.

2) a) Deposit         = £30
      Monthly payments = **£240**
      Total HP cost    = **£270**

   b) £50

3) a) £25    b) £40

### Challenge

1) Holiday Heaven £2550, Sunny Seas £2660 and Beach Bound £2700

2) Beach Bound

## 8 Time

### 8.1 Investigating how long a journey will take (p.127)

### Before we start

1 hour 15 minutes

### Questions

1) a) 11·29 am    b) 12 am Thursday    c) 4·58 pm

2) 2·05 pm

3) a) 2 hours    b) 1 hour 7 minutes    c) 28 minutes

### Challenge

Answers will vary.

## 8.2 Calculating the duration of activities and events (p.129)

### Before we start

Answers will vary.

### Questions

1) Answers will vary.

2) i) 180 seconds; 3 minutes    ii) 48 hours; 2 days

3) i) 510 minutes; 8.5 hours    ii) 192 hours; 8 days

### Challenge

i) Seconds 14934; 248 minutes, 54 seconds; 4 hours, 8 minutes, 54 seconds

ii) 2758 minutes; 45 hour, 58 minutes; 1 day, 21 hours, 58 minutes

## 8.3 Investigating ways speed, time and distance can be measured (p.132)

### Before we start

a) 320 miles    b) 140 miles    c) 250 miles

### Questions

1) a) 5 hours    b) 4 hours    c) 10 hours    d) 4 hours

2) a) 80 mph    b) 270 kph    c) 70 mph    d) 4 mph

3) a) 225 miles    b) 50 mph    c) five hours

### Challenge

a) 160 m per minute    b) 2·5 minutes

## 8.4 Calculating time accurately using a stopwatch (p.134)

### Before we start

a) True    b) False    c) False    d) False

### Questions

1) Answers will vary.

2) Answers will vary.

3) Answers will vary.

### Challenge

Time elapsed

4 minutes 12 seconds

7 minutes 43 seconds

11 minutes 11 seconds

28 minutes 14 seconds

## 8.5 Converting between units of time (p.136)

### Before we start

a) minutes    b) hours    c) seconds    d) years

### Questions

1) Fortnight – two weeks, 48 hours – two days, 10 years – Decade, Century – 100 years

2) a) 5 minutes    b) 9 minutes
   c) 15 minutes    d) 30 minutes

3) a) 4 hours    b) 8 hours 30 minutes
   c) 6 hours 30 minutes    d) 12 hours 30 minutes

4) a) 4 days    b) 7 days    c) 14 days    d) 30 days

### Challenge

Millisecond – one thousandth of a second

Microsecond – one millionth of a second

Decasecond – a unit of time equal to 10 seconds

Quarter – one fourth of the year

Olympiad – a period of four years

Jubilee – an anniversary of an event (silver jubilee = 25 years, golden jubilee = 50 years … )

Mega-annum – one million years

# 9 Measurement

## 9.1 Estimating and measuring length (p.138)

### Before we start

6·8 cm     130 cm     0·7 cm     1750 cm

### Questions

1) Lounge: 4·84 m × 5·78 m, Bedroom: 4·47 m × 5·82 m, Hall: 7·63 m × 3·99 m, Bathroom: 3·85 m × 2·1 m, Kitchen: 4·63 m × 4·63 m, Dining room: 4·58 m × 4·71 m, Utility room: 1·55 m × 4·77 m, Balcony: 0·98 m

2) a) Australia: 5·31 km,    b) Spain: 4·65 km,
    c) Britain: 5·89 km,    d) Germany: 4·57 km

3) a) 2·24 cm    b) 0·97 cm    c) 1·38 cm
    d) 3·05 cm    e) 4·2 cm    f) 9·31 m

### Challenge

Answers will vary.

## 9.2 Estimating and measuring mass (p.141)

### Before we start

1·2 kg = 1200 g    2·3 kg = 2300 g    3·5 kg = 3500 g

### Questions

1) a) Answers may vary - should be close to:
      ii) 1530 g or 1·53 kg    iii) 1390 g or 1·39 kg
      iv) 410 g or 0·41 kg    v) 4270 g or 4·27 kg
      vi) 2150 g or 2·15 kg

    b) Total weight = 10·03 kg. She will need to remove items to reduce the weight to 10 kg.

2) Answers will vary.

### Challenge

Answers will vary.

## 9.3 Estimating and measuring area (p.143)

### Before we start

112 m²     195 m²

### Questions

1) a) 14·8 cm²    b) 27·2 cm²    c) 35·1 cm²
    d) 32·5 cm²    e) 29·6 cm²

2) Mobile phone screen: 10 cm   Football sticker: 6·3 cm
Tablet screen: 15·5 cm   Classroom window: 2·9 m
Classroom floor: 8·3 m

### Challenge

a) Finlay worked out (3 × 2) + (0·8 × 2) + (3 × 0·6) + (0·8 × 0·6) = 6 + 1·6 + 1·8 + 0·48 = 9·88

b) Orange: 7·05 m²    Blue: 10·08 m²    Green: 7·92 m²

## 9.4 Estimating and measuring capacity (p.146)

### Before we start

a) 1·8 l = 1800 ml    b) 0·6 l = 600 ml    c) 3·1 l = 3100 ml

### Questions

1) a) cranberry juice: 80 ml or 0·08 L, mango juice: 260 ml or 0·26 L, blackcurrant juice: 150 ml or 0·015 L, orange juice: 440 ml or 0·44 L, apple juice: 390 ml or 0·39 L, soda water: 410 ml or 0·41 L.

    b) Fruit punch: 1870 ml or 1·87 L

2) b) 0·07    c) 0·16    d) 0·33    e) 0·85    f) 1·24

### Challenge

Answers will vary.

## 9.5 Estimating imperial measurements (p.149)

### Before we start

Answers will vary.

### Questions

1) Matterhorn: 14 690 ft     Denali: 20 235 ft
Mont Blanc: 15 780 ft     Ben Nevis: 4410 ft
Everest: 29 030 ft     Kilimanjaro: 19 340 ft
Scafell Pike: 3210 ft     Eiger: 13 025 ft

2) Answers may vary but should be near to these answers:
    a) 32 miles    b) 85 miles    c) 93 miles
    d) 32 miles    e) 168 miles    f) 18 miles

### Challenge

a) i) 14 stone; ii) 28 stone; iii) 63 stone; iv) 112 stone; v) 224 stone

b) Answers will vary.

## 9.6 Converting imperial measurements (p.152)

### Before we start

54 inches     80 ounces     2·5 gallons

### Questions

1) a) 5·9 inches    b) 8·2 feet    c) 2·65 ounces
    d) 7·92 pounds    e) 7·86 stone    f) 340·8 ml
    g) 14·2 L    h) 2·2 gallons

2)

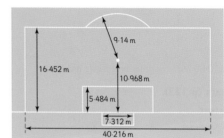

9·14 m = 914 cm       5·484 m = 548·4 cm
16·452 m = 1645·2 cm       7·312 m = 731·2 cm
10·968 m = 1096·8 cm       40·216 m = 4021·6 cm

3) 679·2 g summer fruits       113·2 g madeira cakes
226·4 g caster sugar       120 ml sherry
70·75 g custard powder       568 ml double cream
1·14 L milk       14·2 ml vanilla extract

### Challenge

a) 681 litres     b) 149·7 gallons

## 9.7 Calculating perimeter (rectilinear shapes) (p.155)

### Before we start

Answers will vary but could be rectangles 2 × 3, 3 × 4, 2·5 × 5

### Questions

1) a) 24 cm    b) 20 cm    c) 32 cm    d) 36 cm

2) Trees: 50 m, senior play area: 36 m, infant play area: 32 m, assault course: 24 m, garden: 34 m

### Challenge

Answers will vary.

## 9.8 Calculating area: triangles (p.158)

### Before we start

45 cm² and 64 cm²

### Questions

1) a) 35 cm²    b) 32 cm²    c) 54 cm²

2) Answers will vary.

### Challenge

i) Blue: He can partition the shape into a triangle and a rectangle, calculate their areas and add them together. Red: He can calculate the area of a rectangle, then subtract the area of the triangular section that is missing.

ii) Blue: 1250 cm²       Red: 800 cm²

### 9.9 Measuring area (composite shapes) (p.160)

**Before we start**

Answers will vary but might be 2 cm × 3 cm   3 cm × 3 cm   3 cm × 4 cm

**Questions**

1) Trees: 51 m², senior play area: 65 m², infant play area: 44 m², assault course: 22 m², garden: 88 m²

2) Blue - 26 cm²         Yellow - 21 cm²
   Red - 35 cm²          Green - 24 cm²

**Challenge**

i) The shape can be made into a triangle and a rectangle. Calculate their areas and add them together.

ii) 50 cm²

iii) Blue: 26 cm²         Purple: 48 cm²

### 9.10 Calculating volume (p.163)

**Before we start**

12 cm³    15 cm³

**Questions**

1) a) 648 cm³        b) 2500 m³        c) 3000 cm³
   d) 729 m³         e) 2250 mm³

2) a) i) 1000 cm³    ii) 1200 cm³        iii) 144 000 cm³
   b) 120 maths textbooks        c) 144 dictionaries

**Challenge**

i) 10 cm³    80 cm³    640 cm³

ii) 5120 cm³    40 960 cm³

### 9.11 Volume: composite shapes (p.165)

**Before we start**

64 cm³    105 cm³

**Questions**

1) a) 72 cm³         b) 480 m³
   c) 2400 cm³        d) 320 m³

2) a) A and E        b) B and D        c) B and C
   d) A and B        e) B, C and D

**Challenge**

i) Green: 300 m³    Purple: 300 cm³

ii) Answers will vary.

### 9.12 Capacity problems (p.167)

**Before we start**

192 cm³    576 cm³

**Questions**

1) a) 2·5 L        b) 324 ml        c) 3 L
   d) 4·5 L        e) 10 L

2) a) 360 ml        b) 384 ml        c) 3000 ml
   d) 3125 ml        e) 4050 ml

**Challenge**

1st: 9 L        2nd: 10·5 L        3rd: 12·5 L

The second and third tanks are suitable.

When the decorations are added, the tanks can hold the following capacities of water:

2nd: 8·45 litres

3rd: 10·25 litres

Therefore, Nuria should choose the third tank.

## 10 Mathematics, its impact on the world, past, present and future

### 10.1 Mathematical inventions and different number systems (p.170)

**Before we start**

Answers will vary.

**Questions**

1) a)     b)     c)

   d)     e)     f)

   g)     h)

2) Answers will vary.

3) a)         b)         c)

**Challenge**

Answers will vary.

## 11 Patterns and relationships

### 11.1 Applying knowledge of multiples, square numbers and triangular numbers to generate number patterns (p.172)

**Before we start**

a)

| No. of snowmen (s) | 1 | 2 | 3 | 4 | 5 |
|---|---|---|---|---|---|
| No. of buttons (b) | 5 | 10 | 15 | **20** | **25** |

b) Number of buttons = **5** × number of snowmen

c) b = **5** × s         d) 45 buttons

**Questions**

1) a) 12, 16, 20, 24, **28, 32, 36**    Rule: **Add 4**   Multiple: **4**
   b) 27, 36, 45, 54, **63, 72, 81**    Rule: **Add 9**   Multiple: **9**
   c) 32, 40, 48, 56, **64, 72, 80**    Rule: **Add 8**   Multiple: **8**

2)

| No. (n) | 1 | 2 | 3 | 4 | 5 | 6 | 7 |
|---|---|---|---|---|---|---|---|
| Square number (s) | 1 | 4 | 9 | 16 | **25** | **36** | **49** |

   a) square number = **number** × number or **the number** × **itself**
   b) s = **n** × n
   c) 9th s = 9 × 9
       s = **81**
      10th s = 10 × 10 = **100**

3)

| No. (n) | 1 | 2 | 3 | 4 | 5 | 6 | 7 |
|---|---|---|---|---|---|---|---|
| Triangular number (t) | 1 | 3 | 6 | 10 | **15** | **21** | **28** |

   8th = 28     9th = 36

**Challenge**

b) Finlay will need 55 pom poms

c) Finlay will need 120 pom poms

## 12 Expressions and equations

### 12.1 Solving equations with inequalities (p.174)

**Before we start**

1) a) 5        b) 4

2) a) 7 sweets        b) 3 pupils

**Questions**

1) a) Two squares and two circles is the same as two squares, one circle and two triangles. This means one circle is the same as two triangles.

   b) Two squares, three circles and one triangle is the same as three circles and four triangles. This means two squares equals three triangles.

c) Four triangles, one circle and two squares is the same as five triangles and two squares. This means one circle equals one triangle.

2) a) 6     b) 2     c) 4

3) Various

## Challenge

16

# 13 2D shapes and 3D objects

## 13.1 Describing and sorting triangles (p.177)

### Before we start

It is not a triangle because although it has three straight sides, it has round vertices.

### Questions

1)

| Scalene | Isosceles | Equilateral |
|---|---|---|
| A, D, G, H | C, E, I | B, F |

2) $a = 70°$, $b = 40°$. Isosceles.

$c = 46°$, $d = 110°$. Scalene.

$e = 56°$, $f = 56°$. Isosceles

$g = 60°$, $h = 60°$. Equilateral

3) Answers will vary.

### Challenge

a) There are infinitely many scalene triangles with sides of 5 cm and 8 cm.

b) There are two isosceles triangles:

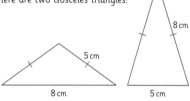

c) There are no equilateral triangles.

d) Where two sides are 5 cm, there are no scalene triangles. There is one equilateral triangle. There are many isosceles triangles, ranging from very flat to very narrow. For example:

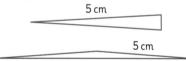

## 13.2 Drawing 2D shapes (p.179)

### Before we start

B and F

### Questions

1) Answers will vary.

2) a) regular octagon     b) irregular hexagon

c) parallelogram     d) equilateral triangle

3) Answers will vary.

### Challenge

a)      b)

## 13.3 Making representations of 3D objects (p.181)

### Before we start

Isla is right. A cylinder is not a prism as it does not have flat faces.

### Questions

1)

2) a) Suitable models should be made.

b)

| Skeletal model | Number of | | |
|---|---|---|---|
| | long straws | short straws | blobs of sticky putty |
| A | 0 | 12 | 8 |
| B | 4 | 4 | 5 |
| C | 3 | 6 | 6 |
| D | 5 | 10 | 10 |

3) a) A C D     b) B     c) A D

## Challenge

a)

| 3D solid | Number of | | | |
|---|---|---|---|---|
| | faces (F) | vertices (V) | edges (E) | F + V |
| cube | 6 | 8 | 12 | 14 |
| hexagonal prism | 8 | 12 | 18 | 20 |
| square-based pyramid | 5 | 5 | 8 | 10 |
| pentagonal prism | 7 | 10 | 15 | 17 |
| triangular prism | 5 | 6 | 9 | 11 |
| pentagonal-based pyramid | 6 | 6 | 10 | 12 |

b) No, because a cone does not have straight edges.

## 13.4 Nets of prisms (p.183)

### Before we start

Nuria is right. The diagram does not show a net of a cube, as it has seven faces.

### Questions

1) A - pentagon-based pyramid; B - cone; C - triangle-based pyramid; D - (short) triangular prism; E - square-based pyramid; F - (long) triangular prism

2) a) A – hexagonal prism; B – cylinder; D – pentagonal prism

b) C is not a prism. The width of the rectangular faces does not correspond with the length of the triangular sides, and the vertices of the triangular faces will not correspond when folded.

3)

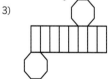

or similar

## Challenge

There are 10 possible nets not including reflections.

# 14 Angles, symmetry and transformation

## 14.1 Drawing triangles (p.185)

### Before we start

Isla is not correct. This can be proven using an example:

### Questions

1) Accurately drawn triangles. Missing angles are

a) 34°     b) 60°     c) 66°     d) 18°

2) Accurately drawn triangles.

a) Angle C is 90°     b) Angle C is 57°

c) BC = 6 cm

Angle B is 33°

Angle C is 114°

3) Answers will vary.

| Length of AB | Angle B | Angle C |
|---|---|---|
| 6 cm | 40° | 50° |
| 7 cm | 36° | 54° |
| 8 cm | 32° | 58° |
| 9 cm | 30° | 60° |
| 10 cm | 27° | 63° |
| 11 cm | 25° | 65° |
| 12 cm | 23° | 67° |

## 14.2 Reflex angles (p.187)

### Before we start
Smallest whole-number reflex angle is 181° and largest is 359°. Pupils may suggest fractional degrees, e.g. 359·9°. The numbers must be between 180° and 360°.

### Questions
1) Answers may vary slightly but should be near to the following:
   a) 344°   b) 267°   c) 193°
   d) 228°   e) 340°
2) a) 173°   b) 49°   c) 114°   d) 91°
3) Answers should be accurately drawn with the reflex angle marked.

### Challenge
Answers will vary.

## 14.3 Finding missing angles (p.189)

### Before we start
Amman is right. Complementary angles add to 90° but there are three angles here. He will need to measure two of them, add them together and subtract from 90° to find the third.

### Questions
1) A = 214°           C = 19°
   B = 121°           D = E = 140°
2) a) A = 163°, B = 156°     b) A = 30°, B = 145°
   c) A = 106°, B = 104°     d) A = 222°, B = 103°
3) A = 160°   B = 55°   C = 31°   D = 31°   E = 239°

### Challenge
A = 96°           C = 130°
B = 100°          D = 34°

## 14.4 Using bearings 1 (p.191)

### Before we start
Finlay is facing south-west.

### Questions
1) Bearings should be accurate to within ± 2°
   1·2 km on a bearing of 065°
   1·4 km on a bearing of 202°
   3 km on a bearing of 070°
   1·5 km on a bearing of 273°
2) 250 km on a bearing of 130°
   150 km on a bearing of 097°
   200 km on a bearing of 048°

### Challenge
Bearings should be accurate to within ± 2°.
Nuria's route is 1·5 km on a bearing of 229°.
Amman's route is 1·7 km on a bearing of 200°, 1·3 km on a bearing of 180°, 1 km on a bearing of 104°.

## 14.5 Using bearings 2 (p.193)

### Before we start
Neither is right. The bearing is 075°. Finlay has measured correctly from north but has not used the three-figure notation. Nuria has measured the angle between the two lines, not the bearing.

### Questions
1) a)
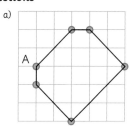

   b) The shape is an irregular hexagon.

2)
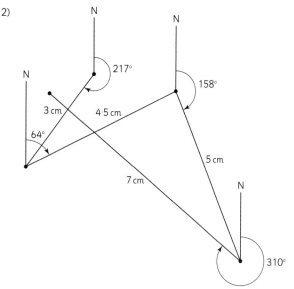

3)
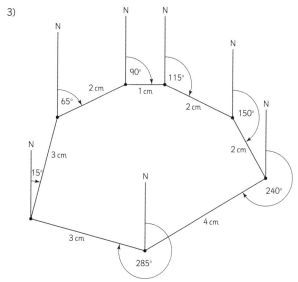

### Challenge
Answers will vary.

## 14.6 Using coordinates (p.195)

### Before we start
Isla will make a rectangle, because (1, 3) and (4, 3) have the same y coordinate and (4, 3) and (4, 5) have the same x coordinate and so on.

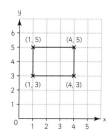

## Questions

1) a)

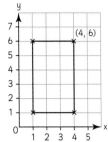

b)

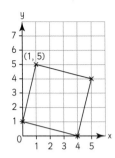

c)

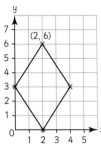

d)

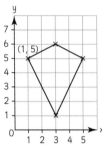

2)  a)  D = (5, 3)                    b)  C = (1, 2)
    c)  B = (3, 2) and D = (7, 3)

3)  a)  D = (2, 12), E = (6, 12)
    b)  D = (5, 1), F = (7, 5), G = (7, 9) and H = (3, 9)
    c)  C = (6, 2), D = (9, 2) and E = (6, 7)

## Challenge

Label each rectangle ABCD and number them 1, 2, 3 …

| Rectangle | Coordinates of A |
|-----------|------------------|
| 1 | (1, 2) |
| 2 | (3, 4) |
| 3 | (5, 6) |
| 4 | (7, 8) |
| ... | ... |
| 10 | (19, 20) |

To find any vertex A, the y coordinate is twice the rectangle number and the x coordinate is one less than the y coordinate. Therefore, for the 10th rectangle A = (19, 20).

Vertex B is A + (0, 2). So, the 10th rectangle, B = (19, 22)
Vertex C is B + (4, 0). So, the 10th rectangle, C = (23, 22)
Vertex D is A + (4, 0). So, the 10th rectangle, D = (23, 20)

## 14.7 Symmetry 1 (p.198)

### Before we start

### Questions

1) a)  b)  c)  d)  e)  f)

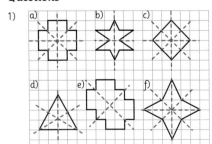

2) One or more diagonal line of symmetry      Vertical *and* horizontal lines of symmetry

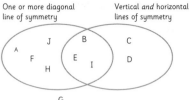

## Challenge

i)

| | Name of shape | Number of lines of symmetry |
|---|---------------|------------------------------|
| A | Square | 4 |
| B | Pentagon | 5 |
| C | Hexagon | 6 |
| D | Heptagon | 7 |
| E | Octagon | 8 |

ii)  F is a nonagon and has 9 lines of symmetry.
     G is a decagon and has 10 lines of symmetry.

iii) A 100-sided regular polygon will have 100 lines of symmetry.

iv)  A 100-sided figure is called a hectogon or a centagon.

## 14.8 Symmetry 2 (p.201)

### Before we start

1)          2)

### Questions

1)

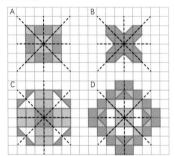

2)

## Challenge

Answers will vary.

## 14.9 Using scale (p.203)

### Before we start

It is about 40 km from Dundee to Perth.

### Questions

1)

| Cities | Distance on map in cm |
|--------|------------------------|
| a)  Rome – Naples | 10 |
| b)  Genoa – Milan | 6 |
| c)  Turin – Milan | 6·5 |
| d)  Florence – Pisa | 3·5 |
| e)  Parma – Modena | 2·5 |

Check the lengths of lines drawn.

2) a) i) 9 cm   ii) 5·2 cm   iii) 4·8 cm
    iv) 2·8 cm   v) 3·4 cm

b)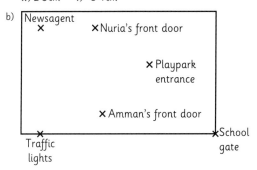

c) Nuria's front door is 3·2 cm from the playpark entrance. In real life this is 80 m.

3) a) 10 cm diameter circle.

b) Tycho crater: 1 cm north and 0·5 cm west
Copernicus crater: 5 cm north then 2·5 cm west.
Plato crater: 8·5 cm north then 1·6 cm west.

c)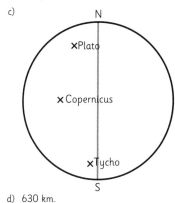

d) 630 km.

## Challenge
Answers will vary.

# 15 Data handling and analysis

## 15.1 Sampling (p.206)

### Before we start
Answers will vary.

### Questions

1) b) This is a whole population because all the staff are asked.
c) This is a whole population because the pupils were asked over a whole week. It is possible that one or two pupils may not have been present to give their opinion, but not enough to make a difference to the data.
d) This is a sample because not everyone who watches the programme will phone in.

2)

| | Big question | Whole population | Sample |
|---|---|---|---|
| b) | What newspapers do the parents of pupils in our school read? | All the parents of all the pupils in our school. | The parents of eight (nine, ten) pupils from each class or other suitable sample. |
| c) | How do the people who work in the local shops travel to work? | All the people who work in the local shops. | One or two people from each shop or other suitable sample. |
| d) | What do librarians in Scotland like to read in their spare time? | All the librarians in Scotland. | One 100 randomly selected librarians or other suitable sample. |

## Challenge

1) Only covers one day.
2) Only those who went to the shop.

## 15.2 Interpreting graphs (p.208)

### Before we start
c) is most suitable as the increments are small enough to see the difference between the data.

### Questions

1) a) misleading  b) accurate  c) accurate  d) misleading

2) a) Graph (i) is misleading because the scale does not start at zero, and the increments are narrow. This exaggerates the difference between the two sets of data.
b) The Herald, because they want their readers to think it is much more popular. People are more likely to read a popular newspaper.

3) a)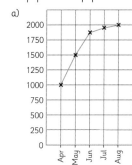
b) The growth looks much steeper.
c) The salesman wants his audience to think that the rival company are not selling as well as they really are. This will make his company look better.

## Challenge

a)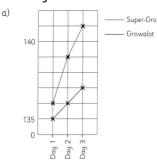
b) She is paid to conduct her research by the company that makes Super-Gro, so she has an interest in making them look good. She invented Super-Gro, so she will want it to be better.

## 15.3 Understanding sampling bias (p.212)

### Before we start
The whole population is all the people on their street. A sample could be every third house on both sides of the street. If the street is really long, they could ask every 10th house.

### Questions

1) a) unbiased    b) biased

2) a) Will give a good picture of the whole population, as the sample is random.
b) Will not give a good picture, because a lot of people cannot go shopping on Tuesdays. Better to take a smaller, random sample on every day of the week.
c) Will not give a good picture as my sister and her friends will probably like the same things, because they are friends. The sample is not random.

3) a) 1) Their sample is biased – they have only asked children who are more likely to watch the news.
2) The survey is not random.
3) They have not asked any adults.

b) They could ask every third (or fourth, or fifth) person on the register in every class, and every third (or fourth, etc.) adult on a list.

## Challenge

Answers will vary.

### 15.4 Pie charts (p.214)

**Before we start**

76·67% attended.

### Questions

1) a)

| Fruit | Number of pieces sold | Number of degrees | Percentage of the total |
|-------|----------------------|-------------------|------------------------|
| Apple | 126 | 126 | 35 |
| Melon | 108 | 108 | 30 |
| Pear | 72 | 72 | 20 |
| Satsuma | 54 | 54 | 15 |
| Total | 360 | 360 | 100 |

b)

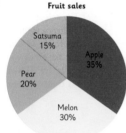

**Fruit sales**

c) Roughly as many apples were sold as satsumas and pears together.

About a third of all fruit sold were apples and about another third was melon.

2) a)

| Job | Number of people | Degrees | Percentage |
|-----|-----------------|---------|------------|
| Ground staff | 864 | 234 | 60 |
| Passport inspector | 72 | 18 | 5 |
| Pilot | 216 | 54 | 15 |
| Flight attendant | 288 | 72 | 20 |
| Total | 1440 | 360 | 100 |

b)

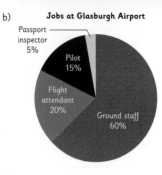

**Jobs at Glasburgh Airport**

Possible answers include: There are more ground staff than all other jobs together.

There are almost the same number of pilots as flight attendants.

About a third of people are working on aircraft (pilots and flight attendants).

## Challenge

a) It looks as though scarves sold the most, then jewellery, then shoes, and bags sold the least.

b)

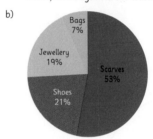

c) The 3D chart is hard to read as there are no numbers.

It is difficult to order the departments as you cannot measure the sections.

## 16 Ideas of chance and uncertainty

### 16.1 Predicting and explaining the outcomes of simple chance situations and experiments (p.217)

**Before we start**

a) 100%    b) - d) Answers will vary. Discuss the opportunity and justify your answers.

### Questions

1) a) $\frac{1}{6}$    b) $\frac{1}{3}$    2) a) $\frac{1}{2}$    b) $\frac{1}{6}$

3) a) $\frac{1}{3}$    b) $\frac{1}{6}$    c) 0

### Challenge

1) 10    2) 10    3) 20    4) 40

© 2019 Leckie

001/27022019

10 9

The authors assert their moral right to be identified as the authors for this work.

All rights reserved. No part of this publication may be reproduced, stored in a retrieval system, or transmitted in any form or by any means, electronic, mechanical, photocopying, recording or otherwise, without the prior written permission of the Publisher or a licence permitting restricted copying in the United Kingdom issued by the Copyright Licensing Agency Ltd., 5th Floor, Shackleton House, 4 Battle Bridge Lane, London SE1 2HX

ISBN 9780008314002

Published by
Leckie
An imprint of HarperCollinsPublishers
Westerhill Road, Bishopbriggs, Glasgow, G64 2QT
T: 0844 576 8126    F: 0844 576 8131
leckiescotland@harpercollins.co.uk    www.leckiescotland.co.uk

HarperCollins Publishers
Macken House, 39/40 Mayor Street Upper, Dublin 1, D01 C9W8, Ireland

Publisher: Fiona McGlade
Managing editor: Craig Balfour
Project editors: Rachel Allegro and Alison James

Special thanks
Answer checking: Rodger Alderson
Copy editor: Louise Robb
Cover design: Ink Tank
Layout and illustration: Jouve
Proofreader: Dylan Hamilton

A CIP Catalogue record for this book is available from the British Library.
Acknowledgements

P48a © josefkubes / Shutterstock.com; P48b © T.W. van Urk / Shutterstock.com; P48c © Morag Fleming / Shutterstock.com;  P48d © Anton_Ivanov / Shutterstock.com
Other images © Shutterstock.com

Whilst every effort has been made to trace the copyright holders, in cases where this has been unsuccessful, or if any have inadvertently been overlooked, the Publishers would gladly receive any information enabling them to rectify any error or omission at the first opportunity.

Printed in the United Kingdom by Martins the Printers

MIX
Paper | Supporting responsible forestry
FSC™ C007454

This book is produced from independently certified FSC™ paper to ensure responsible forest management.

For more information visit: www.harpercollins.co.uk/green